AS UNIT 1

STUDENT GUIDE

CCEA

Geography
Physical geography

Tim Manson and Alistair Hamill

HODDER
EDUCATION
AN HACHETTE UK COMPANY

Hodder Education, an Hachette Company, Carmelite House, 50 Victoria Embankment, London, EC4Y 0DZ

Orders

Hachette UK Distribution, Hely Hutchinson Centre, Milton Road, Didcot, Oxfordshire, OX11 7HH

tel: 01235 827827

e-mail: education@hachette.co.uk

Lines are open 9.00 a.m.–5.00 p.m., Monday to Friday. You can also order through the Hodder Education website: www.hoddereducation.co.uk

ISBN 978-1-4718-6309-7

First printed 2016

Impression number 8

Year 2023

This Guide has been written specifically to support students preparing for the CCEA AS Unit 1 Geography examinations. The content has been neither approved nor endorsed by CCEA and remains the sole responsibility of the author.

Cover photo: Alistair Hamill

Typeset by Integra Software Services Pvt. Ltd., Pondicherry, India

Printed and bound by CPI Group (UK)Ltd Croydon CR0 4YY

Hachette UK's policy is to use papers that are natural, renewable and recyclable products and made from wood grown in well-managed forests and other controlled sources. The logging and manufacturing processes are expected to conform to the environmental regulations of the country of origin.

Contents

■ Getting the most from this book

Exam tips

Advice on key points in the text to help you learn and recall content, avoid pitfalls, and polish your exam technique in order to boost your grade.

Knowledge check

Rapid-fire questions throughout the Content Guidance section to check your understanding.

Knowledge check answers

1 Turn to the back of the book for the Knowledge check answers.

Summaries

■ Each core topic is rounded off by a bullet-list summary for quick-check reference of what you need to know.

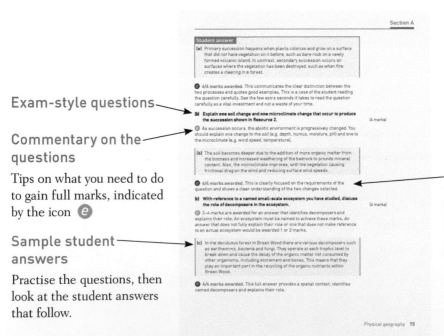

Exam-style questions

Commentary on the questions

Tips on what you need to do to gain full marks, indicated by the icon ⓔ

Sample student answers

Practise the questions, then look at the student answers that follow.

Commentary on sample student answers

Read the comments (preceded by the icon ⓔ) showing how many marks each answer would be awarded in the exam and exactly where marks are gained or lost.

■About this book

Much of the knowledge and understanding needed for AS geography builds on what you have learned for GCSE geography, but with an added focus on geographical skills and techniques, and concepts. This guide offers advice for the effective revision of **AS Unit 1: Physical geography**, which all students need to complete.

The AS 1 external exam paper tests your knowledge and application of aspects of physical geography with a particular focus on fluvial environments, the human impact on ecosystems and weather and climate. The exam lasts 1 hour 15 minutes. The unit makes up 40% of the AS award or 16% of the final A-level grade.

To be successful in this unit you have to understand:
■ the key ideas of the content
■ the nature of the assessment material — by reviewing and practising sample structured questions
■ how to achieve a high level of performance within the exam.

This guide has two sections:

Content Guidance — this summarises some of the key information that you need to know to be able to answer the examination questions with a high degree of accuracy and depth. In particular, the meaning of key terms is made clear and some attention is paid to providing details of case study material in order to help to meet the spatial context requirement within the specification. Students will also benefit from noting the **Exam tips** that provide further help in determining how to learn key aspects of the course. **Knowledge check** questions are designed to help learners to check their depth of knowledge — why not get someone else to ask you these!

Questions & Answers — this includes some sample questions similar in style to those you might expect in the exam. There are sample student responses to these questions as well as detailed analyses, which give further guidance in relation to what exam markers are looking for in order to award top marks.

The best way to use this book is to read through the relevant topic area first before practising the questions. Only refer to the answers and examiner comments after you have attempted the questions.

Content Guidance

■ Topic 1 Rivers

Processes that shape fluvial environments

The drainage basin as a system

A drainage basin is the area of land that is drained by a river and its tributaries. As water falls on the land as precipitation, gravity pulls it downhill and back towards the sea. The boundary of a drainage basin is known as the watershed. This is in the form of a ridge of high land so that any water that falls inside the watershed line will drain through the system. Water that falls outside this boundary will flow through a different drainage system.

Components of a geographical system

The systems approach gives us a very useful framework for analysing and understanding the drainage basin (Figure 1). An open system will have four main elements: **inputs** into and **outputs** from the system (of both energy and matter) and **stores** and **transfers** within the system.

Exam tip

Make sure that you understand the differences between the components of an open system and apply them to the different processes of the drainage basin system.

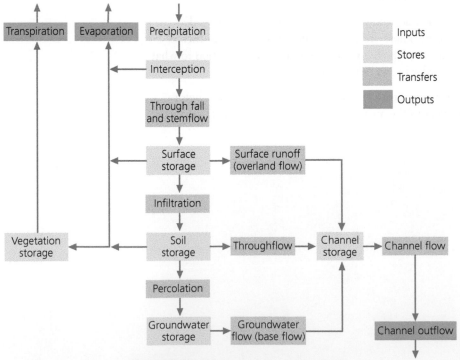

Figure 1 The drainage basin as a system

Components of the drainage basin

1 **Precipitation** (input): water falls from clouds (as drizzle, rain, sleet and snow) towards the surface.

2 **Interception** (store): the precipitation is caught and held for a short time by vegetation before it reaches the soil store. When there is more vegetation, more interception will occur and there is the potential for more evaporation, especially in the summer.

3 **Stem flow and through fall** (transfers): the movement of water from the interception store to the surface store, either by flowing down the stems/branches (stem flow) or dripping off the leaves (through fall).

4 **Surface storage** (store): water is stored temporarily on the surface (e.g. in puddles).

5 (a) **Infiltration** (transfer): water *enters* the soil from the surface store.

(b) **Overland flow/surface runoff** (transfers): water runs over the surface of the land following a rainstorm. This occurs when the soil either becomes **saturated** (because too much rain has fallen and there is no more room in the soil store) or because the rain intensity is too high and **exceeds the infiltration capacity** (i.e. the rate at which water can infiltrate into the soil). It also occurs on **impermeable surfaces**, for example tarmac in urban areas.

6 **Soil storage** (store): water that has infiltrated is stored in the surface layers of the soil before experiencing either throughflow or percolation.

7 **Throughflow** (transfer): water moves downhill through the soil, close to and parallel with the surface, to the river.

8 **Percolation** (transfer): water moves further down into the soil from the soil store to the groundwater store.

9 **Groundwater storage** (store): the permanent store of water in the lower layers of the soil and the bedrock.

10 **Groundwater flow** (transfer): the movement of water from the groundwater store in the lower layers of soil and the bedrock to the river.

11 **Evaporation** (output): water is changed into water vapour from various stores such as interception and surface storage. The main factor affecting the rate of evaporation is temperature.

12 **Transpiration** (output): water vapour is taken from vegetation and plants into the atmosphere. This is affected by factors such as the vegetation type (e.g. deciduous trees lose their leaves in winter to reduce transpiration) and moisture availability.

What factors affect transfers and stores of matter in a drainage basin?

- **Vegetation:** thick vegetation (like forests) can affect the flow of water as it creates more opportunities for interception and for evaporation from the leaves of the trees. Any water that does not directly reach the ground will take a much longer time to go through the system. Soil within dense forests is unlikely to reach infiltration capacity.

Exam tip

When learning these terms, picture the connections between them. Imagine the journey of the raindrop as it transfers between each of the stores.

Exam tip

Top-level answers must use a good range of terms. Be familiar with these terms and practise using them in your answers.

Knowledge check 1

Distinguish between infiltration and percolation.

- **Soil type:** some soils are *porous* (they have big spaces, e.g. sandy soils) and this allows water to infiltrate quickly. Other soils are less porous (e.g. clay soils) and water cannot infiltrate, making overland flow the main transfer.
- **Seasons:** the warmer temperatures of summer encourage more evapotranspiration, thus lowering discharge levels in rivers. If the soil becomes hard-baked it will not allow infiltration. When deciduous leaves fall in the autumn there is less interception. In the winter the ground could be frozen, preventing infiltration.
- **Geology:** if the rock underlying the soil is porous, then there is more percolation and groundwater flow. This reduces the soil store and the amount of water transferred by overland flow.
- **Urban areas:** these areas have less vegetation so there is less interception and less infiltration due to the amount of impermeable surfaces. Water is often channelled directly and quickly into rivers via drains and sewers.
- **Relief:** in steeper drainage basins, less infiltration tends to occur and so *surface runoff* dominates and there is less water in the soil store. Conversely, in lowland areas more infiltration can occur and water tends to remain in the soil store for longer before being transferred via throughflow.
- **How heavy the rain is:** during heavy, intense rainstorms, rain can fall at a faster rate than the soil can accommodate infiltration (the maximum rate of infiltration is called the **infiltration capacity**). If this occurs, overland flow is likely to happen.

Discharge and the storm hydrograph

Discharge is the amount of water that passes a particular point in the river per second. It is measured in cubic metres of water per second (cumecs) and is calculated by multiplying velocity by cross-sectional area.

Discharge in a river is not constant and rises and falls over time, especially in response to a rainstorm. These changes in flow can be plotted on a time/discharge graph called a **storm hydrograph** (Figure 2).

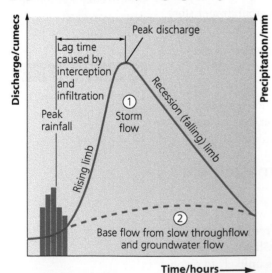

Key

① Runoff from direct overland flow reaches the stream very rapidly and leads to a rapid rise in discharge. Throughflow is slower than overland flow, but nevertheless it contributes to the rising limb too.

② Base flow rises very slowly over many hours after the storm event, because it takes time for it to percolate and make its way via groundwater flow to the river channel.

Figure 2 Storm hydrograph

Exam tip

If an exam question asks for impacts of these factors on both stores and transfers, make sure both are clearly and explicitly covered in your answer. Questions can also explore what happens to stores/transfers if land use changes, for example if a rural area becomes urbanised.

Exam tip

Sediment is also transferred through drainage basins. See page 11 for more information on this.

Knowledge check 2

How do stores and transfers vary throughout the year in a deciduous forest in the UK?

The hydrograph has two main elements:

- **base flow** is the 'background' flow contributed slowly and steadily by groundwater flow
- **storm flow** is the additional flow contributed by precipitation.

Note also the following terms: **peak discharge** (the maximum discharge level), **lag time** (the time difference between peak precipitation and peak discharge), **rising limb** and **falling (recessional) limb**.

As the rainstorm begins, the discharge rises very slowly. There is little initial change during the rainstorm, as most rain does not fall directly into the river. So it experiences interception, infiltration, throughflow and so on, all of which take time.

Later on, the **rising limb** of the discharge rises steeply towards its **peak discharge**. This is because the storm water has now started to reach the channel after experiencing infiltration and throughflow. If the rain is intense and/or if the soil becomes saturated, water arrives via overland flow too.

After reaching its peak, the **falling limb** falls more slowly as the main flow now is throughflow, which is slower than overland flow.

Finally, the discharge returns to its original level. The influence of the storm flow has now passed and the discharge goes back to **base flow**, which is the steady feed of water into the river via groundwater flow.

Factors affecting discharge and the storm hydrograph

The hydrograph in Figure 2 shows a model response (i.e. a typical or average response) to one single rainstorm. In reality, there are many factors that can modify this model response and affect the shape of the hydrograph. We will consider two contrasting types (Figure 3): flashy hydrographs (which have a short lag time and a high peak discharge) and flat hydrographs (with longer lag times and lower peak discharges).

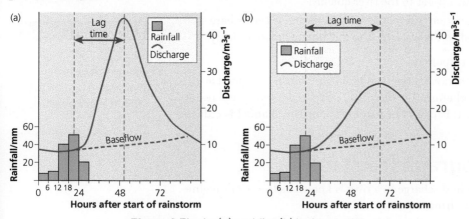

Figure 3 Flashy (a) and flat (b) hydrographs

Nature of the storm

- **Intensity and length of precipitation:** more intense rain causes the soil to become saturated and/or the infiltration capacity to be exceeded, so more overland flow occurs. A similar situation can occur if the rain is prolonged and the soil store becomes filled. This produces the shorter lag time and higher peak discharge of the flashy hydrograph.

Nature of the drainage basin

- **Basin size:** in smaller basins, the precipitation has less distance to travel before it reaches the mouth, so the hydrograph will be shorter and steeper.
- **Basin shape:** shorter, more rounded basins are more likely to be flashy, as the water from the basin tends to arrive more quickly at the mouth. In a longer, thinner basin, the water that falls near the source has much further to travel and so it will produce a flatter hydrograph.
- **Basin relief:** in steeper basins, under the influence of gravity, water will make its way to the mouth more quickly.
- **Soil type:** clay soils have much smaller pore spaces and so do not allow for much infiltration. As a result, overland flow is more likely and so the water reaches the channel quickly. The opposite is the case for sandy soils.
- **Geology:** some rocks, such as basalt, are impermeable and less infiltration occurs. This produces flashier hydrographs. More permeable rocks, such as chalk, allow infiltration and produce flatter hydrographs.
- **Drainage density** (number of streams per km^2): drainage densities are higher in areas with clay soils. As the water does not have to travel far to get to a channel, it travels to the mouth more quickly. This tends to produce shorter lag times and higher peaks.

Land use

Some land uses produce flashy responses.

- **Urban areas:** the impermeable surfaces increase runoff and the drains and sewers are designed to take the surface water to the river quickly.
- **Ploughed fields:** where vegetation is removed for agriculture, it leaves bare soil. This reduces interception and so the water gets to the channel more quickly.

Some land uses produce flatter responses.

- **Afforestation:** this increases interception and thus slows down the speed at which the water reaches the channel. Furthermore, increased interception results in more evaporation, so the total amount of water reaching the channel is reduced, lowering the peak discharge.

The annual hydrograph (regime)

The pattern of annual variation in discharge in a river is known as the river's regime. The main factor controlling it is **climate**.

- **Precipitation** — the total amount of rainfall affects a river's discharge, but so does its timing. Some climates experience **monsoon rainfall** (which means that most rain falls in the summer months giving a large peak in river discharges at this time e.g. River Indus, Pakistan, see page 26). In contrast, in the UK, rain tends to

Source Where drops of water join to start a river.

Mouth Where the river flows into the sea.

Exam tip

Remember to connect clearly and explicitly each of these factors with how they affect: (1) the **lag time** and (2) the **peak discharge**.

Knowledge check 3

How might increased urbanisation in a drainage basin affect lag time and peak discharge?

be distributed more evenly throughout the year and so there is less annual variation in river regime.

■ **Temperature** — the higher the temperature, the greater the rates of evaporation. This reduces the amount of water in a river channel, making discharge peaks lower than they would otherwise have been. In the UK, this tends to reduce discharge levels in the summer months (e.g. River Severn).

■ **Snowmelt** — linked to temperature, this can have a significant impact on rivers that have large amounts of snow and ice in their drainage basins. For example, the Indus River in Pakistan sees its discharge rise in late spring/early summer as a result of snow melting in the Himalayas (see page 26).

River processes (erosion, transportation and deposition)

Erosional processes

Erosion is the wearing away of the bed and banks in the river. There are four main erosional processes.

■ **Abrasion/corrasion** is the most effective form of river erosion and occurs when the river uses its load to erode the bed and banks by scraping and scouring. It is particularly effective at times of higher discharge when the river has enough energy to transport larger particles (see the Hjulström curve on page 12).

■ **Hydraulic action** occurs due to the physical force of the water against the bed and banks. On the outside bends of meanders, for instance, the currents push water into cracks, causing pressure that leads to erosion. Hydraulic action is more effective in rapids and at waterfalls. It tends to move unconsolidated sands and gravels on the riverbed.

■ **Corrosion/solution** is the dissolving of soluble materials in the bed and banks by weak acids in river water. This is a chemical reaction rather than a physical process and so is not dependent on the energy levels in the river. It is most effective in rocks containing carbonates, such as limestone.

■ **Attrition** occurs when the load particles come into contact with other load particles and the bed and banks. As a result the rough edges are smoothed and the particles become smaller and more rounded (this is particularly noticeable as you move downstream).

Erosional processes can occur in two directions:

■ **vertical**, creating river valleys — this is more common in the upper course of the river.

■ **lateral** (horizontal) as meandering rivers widen floodplains.

Transportation processes

Any energy not lost by the river by friction can be used to transport sediment. There are four types of transportation.

■ **Suspension:** the smaller particles of clay, silt and sand can be carried along by the turbulence of the river. This tends to be the most effective form of transportation and it explains why rivers in their lower course tend to be brown in colour.

Knowledge check 4

Which of the erosional processes are affected by changes in river energy levels as discharge increases or decreases?

- **Solution:** the material eroded by corrosion is carried along and dissolved in the water. This form of transportation can be significant in limestone areas, but tends to be less important in other areas.
- **Saltation:** the smaller bedload, such as pebbles and gravel, can be bounced along the riverbed by turbulence during times of higher discharge.
- **Traction:** the largest boulders in the river can be rolled along the riverbed during times of very high discharge.

Exam tip

In the exam, if you need to expand on your explanation of deposition, refer to the details of how deposition occurs in the three features mentioned here.

Deposition processes

When river energy drops, deposition occurs in various places in the river such as the inside bends of **meanders** (these are called **point bar deposits**, see page 14), on **floodplains** as a river overflows its banks (see page 16) and in **deltas** (see page 17).

The Hjulström curve

The Hjulström curve (Figure 4) shows the velocity needed to erode, transport or deposit different sized particles.

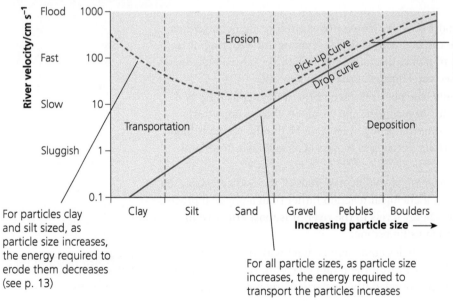

Figure 4 The Hjulström curve

The Hjulström curve shows two lines:

- the **critical erosion velocity/pick-up curve** (i.e. the speed needed to pick up the particles)
- the **critical deposition velocity/drop curve** (i.e. the speed at which there is no longer enough energy to transport the particles).

To understand the graph, we start with this key principle: **larger particles with greater mass require more energy both to erode and transport them**. For example, gravel will be eroded at velocities around $5\,\mathrm{m\,s^{-1}}$, whereas larger boulders require velocities closer to $50\,\mathrm{m\,s^{-1}}$ before they will be eroded. This is true for most of the graph: as particle size increases, the amount of energy required to erode and transport sediment increases (a positive correlation).

However, there is one section of the graph where this positive correlation does not apply. For particles of silt or clay, as particle size decreases, the amount of energy required to erode them actually increases. For example, velocities need to be up around $50\,\mathrm{m\,s^{-1}}$ again before the smallest clay particles will be eroded. There are two main reasons for this.

- Clay particles are **cohesive** (they stick together) and so need more energy to separate them one from another. Once separated, however, they are so small that they are very easily transported.
- Clay particles form **smooth surfaces** when packed together. This means that, as the river water flows over them, there is less turbulent flow, and so fewer eddies in the water that could scour off particles from the surface. In contrast, particles that are sand sized and larger encourage these eddies.

In all cases, the velocities required to erode are greater than those required to transport. However, for particles gravel sized and bigger the difference between the erosional and depositional velocities is quite small. This means that larger particles are deposited soon after they have been eroded. However, for the smallest particles, river velocities have to be very low for deposition to occur. As a result, silts and clays tend to be deposited after flooding on floodplains as the water infiltrates into the soil. They can also be deposited in deltas and estuaries, aided by the process of flocculation (see page 17).

Formation of river landforms

Waterfalls

Waterfalls form where bands of harder rock (such as basalt) overlay bands of softer rock (such as chalk). As the river flows from the harder to the softer rock, the softer rock tends to be eroded more quickly. This initially forms a **step** where the two rock types meet. This causes the water to fall vertically, losing its contact with the bed and increasing its speed due to the drop in friction.

Over time, erosion (especially hydraulic action and corrasion) is concentrated at the base of the waterfall. This excavates the softer rock vertically downwards to create a **plunge pool** and undercuts the harder rock laterally upstream to form a **notch**. As the notch grows in size, the overhanging harder rock above it collapses as it is now unsupported.

By this process, the waterfall tends to migrate upstream, leaving behind a steep-sided gorge (Figure 5). One of the most famous waterfalls in the world is Niagara Falls.

Rapids

Rapids are found in steeper sections of the river channel, for example as water flows over a waterfall, or where the water flows over bands of harder rock that are at a slight angle. Both river velocity and river turbulence are increased here, and so erosion rates are higher. An example of rapids is the Sheer Wall Rapids in the Colorado River.

Exam tip

Take time to understand *why* the relationships shown in the Hjulström curve exist, especially the key principles relating to particle *size* and particle *cohesion*.

Exam tip

Make sure you can make *reference to places* for each of the river features.

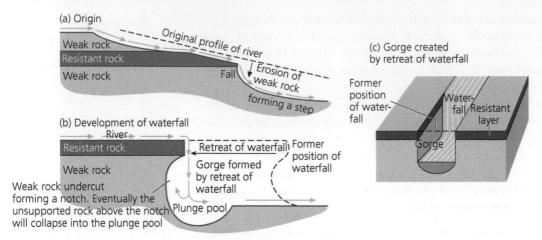

Figure 5 Waterfall formation

Meanders, pools and riffles

Along the riverbeds of rivers carrying particles that are generally sand size and larger, you will often have alternating features called **pools** (deeper areas with smaller particles such as silt) and **riffles** (shallower areas made up of larger particles such as gravel or pebbles). Due to the increased frictional drag caused by the larger particles in the riffles, the maximum velocity flow tends to swing to avoid these areas. This creates side-to-side motion within the water and is the start of the process of meander development.

Once this side-to-side motion is established, it produces variations across the channel. As the maximum velocity flow swings to avoid the riffles, erosion (especially corrasion and hydraulic action) is concentrated on one of the banks. This moves it back laterally, deepening the channel here and causing the undercutting of the riverbank, producing a steep-sided **river cliff**. At the same time, on the other bank where river velocities are lower, deposition occurs, forming a gently sloping bank made up of **point bar deposits**. The slope down towards these deposits is known as a **slip-off slope**. Over time, as erosion continues on the outside of the bend and deposition on the inside, the river becomes more **sinuous** as it moves laterally across the floodplain (Figure 6).

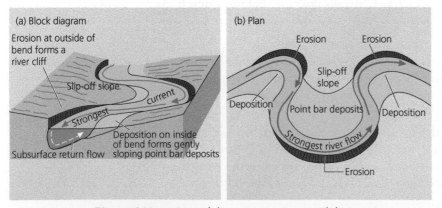

Figure 6 Meander in (a) cross-section and (b) plan view

Exam tip

Questions on river features sometimes require annotated diagrams. You can earn full marks for a clear, fully annotated diagram that incorporates explanations. Practise drawing and annotating all the diagrams in this section. The success criteria for diagrams are accuracy, clarity and detailed annotation.

Knowledge check 5

Describe the cross-sectional shape of a meander.

As well as moving laterally, meanders tend to migrate downstream. This is because the maximum velocity flow does not follow the precise shape of the channel and thus the point of maximum erosion is just downstream of the halfway point of the meander bend. The Mississippi River has many large meanders along many kilometres of its course.

Oxbow lakes

As the meander forming processes continue, the river's sinuosity increases. Over time, continued erosion of the outside of the bend causes the gap in the meander loop to become narrower.

During times of high flow, as the river has more energy and is seeking the most efficient way downstream, it might cut across this gap. This produces a straight channel and thus the maximum velocity flow is now in the middle and deposition occurs at the edges.

Over time, this deposition builds up to cut off the old meander loop, creating the **oxbow lake** (Figure 7). This lake can silt up, forming a crescent-shaped marsh called a **meander scar**. Again, the Mississippi has many examples of oxbow lakes.

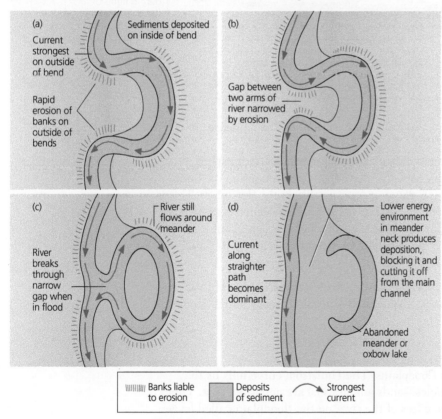

Figure 7 Formation of an oxbow lake

Floodplains

Floodplains are large, flat areas around lowland rivers characterised by large amounts of deposition (known as alluvium).

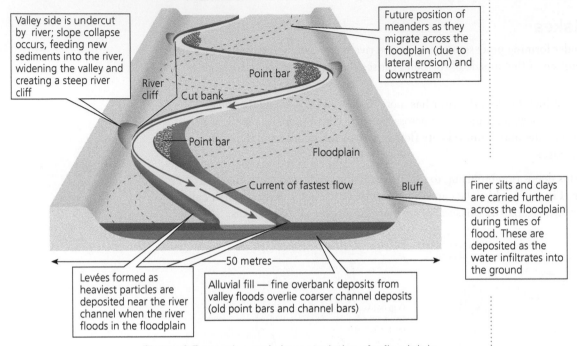

Valley side is undercut by river; slope collapse occurs, feeding new sediments into the river, widening the valley and creating a steep river cliff

Future position of meanders as they migrate across the floodplain (due to lateral erosion) and downstream

Point bar

River cliff

Cut bank

Point bar

Floodplain

Current of fastest flow

Bluff

Finer silts and clays are carried further across the floodplain during times of flood. These are deposited as the water infiltrates into the ground

50 metres

Levées formed as heaviest particles are deposited near the river channel when the river floods in the floodplain

Alluvial fill — fine overbank deposits from valley floods overlie coarser channel deposits (old point bars and channel bars)

Figure 8 Formation and characteristics of a floodplain

The deposition comes from two main sources, the first of which is **migrating meanders**. As the meanders weave laterally, they deposit **point bars** on the inside of the bends as they migrate, leaving alluvium deposits all over the floodplain. This is the main source of deposits on a floodplain. Further deposition occurs as a result of **river flooding**. If the river overflows its banks, it experiences a significant increase in friction (due to contact with the ground and with floodplain vegetation). This lowers the velocity, encouraging deposition.

The meanders also widen the floodplain by lateral erosion at the outside of their bends, often leaving prominent slopes called **bluff lines** at the edge of the floodplain (Figure 8).

Levées

Levées are often found on river floodplains. The Hjulström curve tells us that, as energy drops, the largest particles (sands and gravels) are deposited first. They tend to form low ridges along the edges of the river on the floodplain called levées (see Figure 8). Meanwhile, the finer silts and clays are carried further away from the river across the floodplain and are deposited as the river water slowly infiltrates into the ground over time.

As the levées increase over time, and if there is further deposition on the riverbed, this can result in the river flowing at a higher level than the floodplain. In some places, such as the Mississippi, river engineers have increased the height of the natural levées to increase channel capacity and attempt to reduce the flood risk.

Deltas

Deltas are depositional features extending from the mouth of the river into the sea or a lake.

Factors affecting delta formation

Deltas form under the following conditions.

- **When the rate of deposition exceeds the rate of erosion:** Deposition rates are higher when a river has a large amount of sediment, for example a large river such as the Nile. Erosion rates are lower when the marine environment has a smaller tidal range and weaker currents.
- **Flocculation:** When fresh water mixes with seawater, chemical reactions with the salt cause clay particles to coagulate (stick together). This increases their weight and increases depositional rates.
- **A gentle sea floor gradient:** This increases deposition and aids delta formation.

Delta characteristics

As the sediment leaves the river mouth, it tends to be deposited in three layers: topset beds, foreset beds and bottomset beds (Figure 9a). As the river carries its sediment load out into the open water and the energy levels drop, the Hjulström curve (Figure 4, page 12) tells us that the heavier sands and silts are deposited first while the lighter silts and clays are carried further out.

Delta types

Depending on the overall energy environment, the plan view pattern that can be formed is one of two types.

Arcuate deltas (Figure 9b) grow out from the coastline with a convex outer edge. One of the best examples of this is the delta of the Nile. The deposition at the mouth of a river is usually triangular in shape. This is due to the blocking of the river mouth, which forces the river to break into a number of other channels called **distributaries** at angles to the original course. The distributaries weave back and forth, depositing sediment, which, over time, forms the characteristic fan shape of the arcuate delta. In addition, wave erosion tends to smooth off the outer edge, giving the arcuate delta a more pronounced fan-like shape.

Bird's foot deltas (Figure 9c) occur where delta formation is more river-dominated and less subject to tidal or wave action. The deposition pattern appears more haphazard and the delta can take on a multi-lobed shape that resembles a bird's foot. An example of a river-dominated delta is the Mississippi River delta.

> **Knowledge check 6**
>
> Describe **one** factor that can determine the rate at which a delta forms.

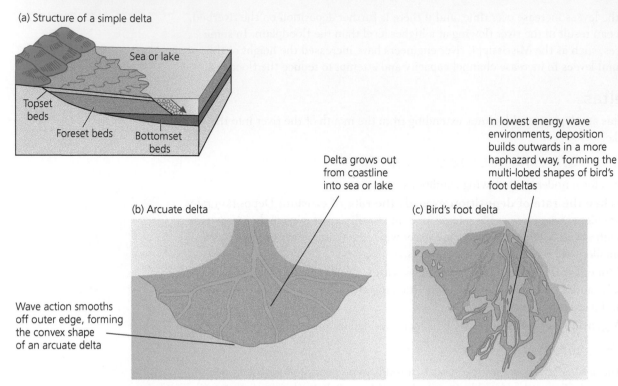

(a) Structure of a simple delta

Sea or lake

Topset beds

Foreset beds

Bottomset beds

(b) Arcuate delta

Delta grows out from coastline into sea or lake

Wave action smooths off outer edge, forming the convex shape of an arcuate delta

(c) Bird's foot delta

In lowest energy wave environments, deposition builds outwards in a more haphazard way, forming the multi-lobed shapes of bird's foot deltas

Figure 9 Deltas in (a) three bed layers, (b) arcuate and (c) bird's foot

Summary

- A drainage basin is an open system consisting of inputs (especially precipitation), stores (such as interception and soil store), transfers (such as overland flow and infiltration) and outputs (including evaporation and transpiration).
- The ways in which the stores and transfers operate can be influenced by a number of factors including vegetation, soil type and seasons. Land use change (such as the change from vegetation to urban land uses) can also affect the stores and transfers.
- The changes in a river's discharge following a storm can be shown in a storm hydrograph, which shows peak discharge and lag time.
- Storm hydrographs can be flashy or flat depending on a range of factors such as the nature of the rainstorm (amount and intensity), characteristics of the drainage basin (basin size, relief, drainage density) or land use (vegetated or urbanised).

- The variations in annual river regimes are determined mostly by climatic factors including amount and timing of precipitation, temperatures (which affect rates of evaporation) and snowmelt.
- Rivers erode, transport and deposit sediment by a variety of processes (corrasion/abrasion, hydraulic action, corrosion and attrition; suspension, solution, saltation and traction).
- The different velocities needed to erode, transport or deposit particles of differing sizes is shown on the Hjulström curve. As particle size increases, the energy needed for erosion and transportation increases. However, for the smallest particles of silt and clay, due to their cohesive nature, more energy is required to erode them, even though they are smaller.
- These river processes form a number of different river features, including: waterfalls, rapids, meanders, pools and riffles, oxbow lakes, levées, floodplains and deltas (arcuate and bird's foot).

Human interactions in fluvial environments

River management

Humans have always been drawn to settle beside rivers due to the various benefits of living there. However, as the last section has shown us, rivers and their floodplains are dynamic. River channels erode banks and deposit sediment; and they meander across floodplains over time. Floodplains themselves experience periodic flooding, as a river discharge varies naturally. All of this means that rivers have been subjected to various forms of management to reduce the impacts of these natural processes on the people living and working nearby. We will examine two different approaches to river management below: **channelisation** and **sustainable and environmentally sensitive solutions**.

Channelisation

Channelisation refers to:

- attempts by humans to modify the size and shape of a river channel
- in order to increase the capacity (how much it can hold) and hydraulic efficiency (how fast it allows water to flow downstream) of the channel.

The methods of channelisation

Re-sectioning

This involves changes to the river's cross-sectional profile, by **widening** and/or **deepening** the river. As a result, the river has:

- a greater **capacity** and so it can hold more water before overflowing its banks onto the floodplain and
- a greater **efficiency** — it allows the water to flow at higher velocities and so water can be removed from this section of the river more quickly, reducing the flood risk in the channelised section.

Exam tip

Note that, although channelisation may indeed reduce the flood risk in the part of the river in which it is being used, it can in fact increase the flood risk downstream of the channelised section.

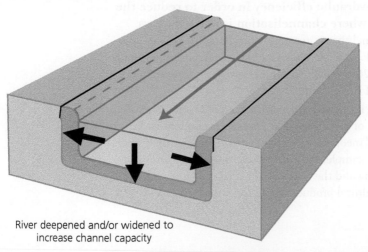

River deepened and/or widened to increase channel capacity

Figure 10 River re-sectioning

Realignment

This involves the **straightening of a river channel**, either by replacing an entire stretch of meandering river with a straighter channel, or creating cut-offs across meander bends. This improves the river's efficiency by increasing river velocity. Velocity is increased, as the removal of meanders results in an increase in the gradient of the river channel.

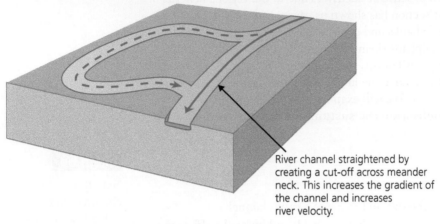

River channel straightened by creating a cut-off across meander neck. This increases the gradient of the channel and increases river velocity.

Figure 11 River realignment

Knowledge check 7

Contrast how re-sectioning and realignment attempt to modify river flow.

Dredging

As part of their natural, dynamic cycle, rivers will deposit material eroded from upstream. Over time, this can cause the channel cross-sectional area to shrink, reducing the capacity of the river. This may require dredging, which is the removal of accumulated silt and sediment from the bed of the river by diggers or pumps. The sediment may also be dislodged from the beds and placed in the main current and allowed to flow further downstream.

The need for channelisation on some rivers

The main reasons for carrying out channelisation include:

■ **to increase river capacity and hydraulic efficiency in order to reduce the impacts of flooding in the area where channelisation is used**. If a river can hold more water, and can channel that water more quickly out of an area, the floodplain around it (and anything located there) is less likely to flood. For example, following the devastating floods of 2014 in the Somerset Levels, the Environment Agency dredged the River Tone and River Parrett.

■ **to improve land drainage in agricultural areas**. Flooding of farmland close to a river can take productive land out of production for weeks or even months during a year, potentially reducing farmers' incomes. By reducing flooding, channelisation can keep this land productive. For example, the River Blackwater Land Drainage Scheme (1984–91) involved dredging and the removal of 13 meanders in order to reduce flooding and increase agricultural productivity in the area.

- **to improve navigation in a river**. Rivers are natural transport routes and vital for economic activity. However, given the size of many modern ships and boats, channels often have to be deepened by dredging and realigned in order to make them more navigable. For example, the Lower Mississippi River has been shortened by around 235 km in the past 90 years by cutting off meander loops.

Channelisation involves hard engineering and so can have some negative impacts associated with it (see below). However, despite the preference for soft engineering approaches to managing rivers today, still there is the need to use channelisation approaches in certain rivers. For example:

- where a river flows through an urban area with development right along the banks, there may not be the space to create safe flood zones beside the river, so it may need to be channelised along these stretches to protect people and property from flooding there. For example, the River Greta is channelised where it runs through Keswick in Cumbria, protecting 141 properties.
- where, for economic reasons, rivers need to be kept navigable for shipping, dredging will have to be used. Ongoing dredging takes place on the Mississippi River, for example, as US Federal law requires the river to have its depth maintained at no less than 3 m, in order to aid navigation.

> **Hard engineering** Typically uses human engineering strategies to try to control the natural processes, rather than to work with them.

Sustainable and environmentally sensitive river management

Why sustainable and environmentally sensitive methods are necessary

The modern approach to river management sees a river as part of a larger connected drainage basin. It favours soft engineering strategies, which use natural drainage basin processes to manage flooding while at the same time enhancing the natural environments of the river channel and floodplain. The main reasons why such approaches are needed are as follows.

- **Due to the limitations of channelisation as a flood control strategy.** Channelisation techniques can be effective at reducing flooding in the area in which they are used. However, they can lead to increased flooding downstream where channelisation is not occurring. When the increased amounts of water moving more quickly move out of the channelised section of river, they are funnelled into a smaller river channel that is less efficient at moving this water on. The result is an increased flood risk.

  In contrast, soft engineering approaches that work *with* natural processes and take a basin-wide approach tend to be more effective at managing the flood risk across the entire drainage basin (see for the techniques used to do this).

- **To restore natural rivers and increase habitat diversity.** One other consequence of channelisation is the reduction of habitat diversity within the river channel itself and across the floodplain. To return rivers to a more natural state, environmentally sensitive river restoration schemes can be used (see Figure 12). These improve habitat diversity and recreation value for people.

> **Soft engineering** These approaches seek to work with natural processes to manage rivers and drainage basins.

■ **Due to the increased threat of flooding resulting from climate change.**
A report published in 2014 suggested that, as a result of climate change, flood damage across Europe will increase four-fold by 2050. This would make floods that currently occur once every 16 years likely to occur once every 10 years by the middle of the century. Given this increasing risk to people, property and the land there is an increased need for sustainable flood management strategies to be implemented.

Methods of sustainable and environmentally sensitive river management

When it is decided to manage a drainage basin using sustainable and environmentally sensitive techniques, the following methods may be used (Figure 12).

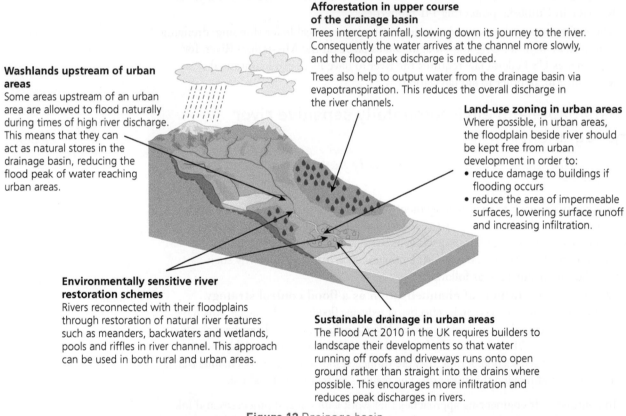

Afforestation in upper course of the drainage basin
Trees intercept rainfall, slowing down its journey to the river. Consequently the water arrives at the channel more slowly, and the flood peak discharge is reduced.

Trees also help to output water from the drainage basin via evapotranspiration. This reduces the overall discharge in the river channels.

Washlands upstream of urban areas
Some areas upstream of an urban area are allowed to flood naturally during times of high river discharge. This means that they can act as natural stores in the drainage basin, reducing the flood peak of water reaching urban areas.

Land-use zoning in urban areas
Where possible, in urban areas, the floodplain beside river should be kept free from urban development in order to:
• reduce damage to buildings if flooding occurs
• reduce the area of impermeable surfaces, lowering surface runoff and increasing infiltration.

Environmentally sensitive river restoration schemes
Rivers reconnected with their floodplains through restoration of natural river features such as meanders, backwaters and wetlands, pools and riffles in river channel. This approach can be used in both rural and urban areas.

Sustainable drainage in urban areas
The Flood Act 2010 in the UK requires builders to landscape their developments so that water running off roofs and driveways runs onto open ground rather than straight into the drains where possible. This encourages more infiltration and reduces peak discharges in rivers.

Figure 12 Drainage basin

River flooding: causes and impacts

Rivers flood naturally. However, human actions can make the scale of those floods greater. In the short term, large-scale flooding has many negative impacts on people (loss of life, damage to homes, economic costs) and property (destruction and damage). However, in the longer term flooding can bring some benefits too, such as improvement to wetland habitats, increased soil moisture for agriculture, and increased fish stocks for food and recreational activities.

Knowledge check 8

'River flooding is entirely negative.' Discuss the extent to which you agree with this statement.

Case study

Causes and effects of flooding in the Mississippi Basin, 2011

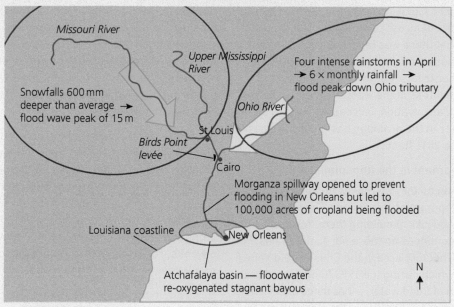

Figure 13 Map of the Mississippi Basin, showing some of the causes and impacts of the 2011 flood

The causes of the 2011 flood

Physical causes

Physical causes were the primary trigger of the devastating floods of 2011 in the Mississippi Basin, in the form of two sets of unusually extreme weather events occurring at similar times in different parts of the drainage basin.

Snowmelt in the Rockies

First, in the upper Mississippi and Missouri **tributaries** to the west (Figure 13), the **snowfalls** of winter 2010/11 were record breaking, with the snow being around 600 mm deeper than average. This resulted in a snowpack over 200% deeper than usual. In addition, a cooler-than-average spring delayed the snowmelt. When snowmelt finally occurred in April, this meant that there was a higher quantity of water entering the river system. This raised the peak discharges and sent a flood pulse down these tributaries that rose to 15 m in Thebes, Illinois (just to the south of the confluence of these two rivers). Furthermore, in May, a year's worth of rain fell in one month. This fed even more

water down the drainage basin into a river system already in flood.

Intense rain in the Ohio tributary basin

At the same time as these events, the Ohio tributary to the east was also having record-breaking weather, but in this case in the form of a series of **four intense rainstorms** in April, which produced six times the average monthly rainfall for the area. This also sent a flood peak down this tributary towards the **confluence** of the Ohio with the Mississippi at Cairo, Illinois.

Human causes

These two physical events alone would have caused flooding. But some human causes increased the scale of the floods.

Deforestation

The first of these was **deforestation**. There has been extensive deforestation in the Mississippi Basin since European settlers began to arrive in the 18th century. At that stage, the Lower Mississippi between Cairo and the delta had 10 million hectares of hardwood and swamp forests. Between 1800 and 1935, 50% of this forest was cleared for agriculture. From 1950 to 1970, a further 3 million hectares was cleared for rice and soybean crops, leaving just 2 million hectares (a total loss of 80% of forest cover since 1800). Fewer trees meant that there was less interception of rain when it fell, causing peak discharges to rise more quickly to higher levels. Furthermore, a lack of trees meant that less of the water was taken up and outputted via transpiration, increasing the overall volume of water in the drainage basin. This increased the scale of the 2011 flood.

Urban development in the floodplains

A second human factor that makes flooding worse is urban development on floodplains. Despite the well-known flood risk of building there, further development continues on these particular plains. For example, an 8-hectare area in the Chesterfield Valley in St Louis — which during the 1993 flood was a rural area under 3m of flood water — has since experienced urban development. It has one of the largest strip malls in the country, along with 28,000 new homes. In the Missouri area, $4.3 billion was spent on urban development on land that was flooded in 1993. The growth of urban areas increased surface run-off, resulting in faster and higher peak discharges, and reduced the storage capacity of the river basin, increasing the total volume of discharge in the river.

The effects of flooding in the Mississippi Basin on people, property and the land

People

The short-term impacts of the flood on people were **negative**. The US Army Corps of Engineers (USACE) estimated that more than 43,000 people felt some effects of the flooding. Despite the huge scale of the flood, the number of fatalities was quite low overall, largely due to the excellent forecasting and evacuation plans. Thirty people lost their lives in total. Most of these deaths (20 people) occurred in the earlier part of the flooding, particularly as drivers ignored road-closed signs and were swept away by the floods.

People were also displaced from their homes due to the flooding. For example, when the USACE blew a 3km hole in the Birds Point levée to protect the city of Cairo from flooding, around 200 residents in New Madrid County were forced to leave their homes. In addition, the USACE opened the Morganza Spillway north of New Orleans, for the first time in 38 years, to lower the flood peak heading for the city. However, the water that flooded the surrounding land put 25,000 people at increased flood risk, including the 12,000 residents of Morgan City, who had to try to protect their homes with sandbags.

People were also affected by the economic impacts of the floods. Farmers in particular were badly hit, with agricultural losses due to flooded land, including crops estimated at $800 million in Mississippi, $500 million in Arkansas and $320 million in Memphis. In addition, the 19 casinos along the Mississippi were closed down for around 6 weeks, resulting in loss of income for the 13,000 employees and $14 million in lost tax revenue.

However, in the longer term, flooding can bring some **benefits to people** in the Mississippi Basin (especially when it is at a lower scale than the 2011 flood). For example, flooding helps maintain the wetlands of the Louisiana coast. This supports over 2,000 diverse commercial aquaculture activities, including fishing of channel catfish and baitfish, along with recreational sports fishing companies. In 2011, $807 million was spent on fishing in the state of Louisiana (making up 37% of all the wildlife related recreational activities in the state). This clearly contributes significantly to the economy of the area.

Additionally, one recent report stated that around 90% of the Mississippi State's drinking water supply comes from the groundwater store within the basin. Groundwater levels have declined on average by 1.5m between 1995 and 2015. Major flooding events like 2011 go some way to reducing the rate of groundwater decline, bringing benefits in terms of water supply to the people in the basin.

Property

As the flood waters headed south, the impacts of the floods on property were **very negative**. The USACE estimated that the total damage to buildings was around $2.8 billion and that more than 21,000 buildings were affected. Some of this damage resulted from human attempts to prevent damage in other, higher value areas. For example, when the Birds Point levée was blown up on 3 May in order to save Cairo, the flood waters spilled out onto the surrounding landscape and 100 homes were flooded as a result. Over $360 million worth of damage was caused to property and infrastructure here.

Further south, in Memphis, the levées that had been built to protect the downtown area worked, but the flood waters instead covered the suburban areas, flooding 1,300 homes.

Further downstream again, as the flood peak moved through Mississippi State from 15–21 May, it caused $1 billion of damage. For example, despite the presence of temporary levées in and around Vicksberg, flood waters impacted more than 2,600 businesses and homes.

Land

The impact of the flooding on the land was **positive**. Normally, the levées built around the river prevent a lot of water and sediment from reaching the wetlands. However, the Morganza Spillway was opened in May to alleviate the flood risk in New Orleans and Baton Rouge. This allowed the flood waters to spill out into the wetland areas of Louisiana and the Atchafalaya Basin. Fresh water flushed out the bayous of stagnant water and brought in fresh, oxygenated waters and nutrient-rich sediment.

The flood waters also deposited much needed sediment into the wetlands, adding a layer up to 12 cm deep of nutrient-rich mud in areas close to the Atchafalaya distributary. In addition, the flood sediment helped to flush out some of the oil that had covered the Louisiana coast following the devastating 2010 Gulf of Mexico oil spill.

In conclusion, flooding in the Mississippi is mostly negative in the short term, especially its impact on property, but it can bring some longer term benefits, especially to the land.

Exam tip

If you are asked to evaluate whether physical or human causes are more responsible for causing flooding, you should say that *extreme physical events* are the main *trigger* of flooding but that *human factors* can make the *scale* of the flooding worse.

Case study

Causes and effects of flooding in the Indus River, Pakistan, 2010

The causes of the 2010 flood

Physical causes

Snowmelt in the Himalayan foothills

The River Indus is fed each year by melt waters from its source in the Himalayas. These tend to cause discharge to rise in June and typically contribute around two-thirds of the annual discharge for the river. However, in 2010, temperatures were higher than average (2010 was then the joint warmest year on record) and this caused additional melting. In particular, some of the glacial lakes in the Himalayas burst adding extra discharge to the river. Of the 2,420 glacial lakes in the Indus Basin, 52 are prone to lake outburst. This factor alone raised discharge levels higher than normal.

Monsoon rains

In addition, 2010 was an exceptional year of rainfall in Pakistan. The climate here is dominated by the monsoon. This causes a wet season from late June to early October that typically affects all but the deserts of the southeast and the mountains of the north of the country. However, the monsoon

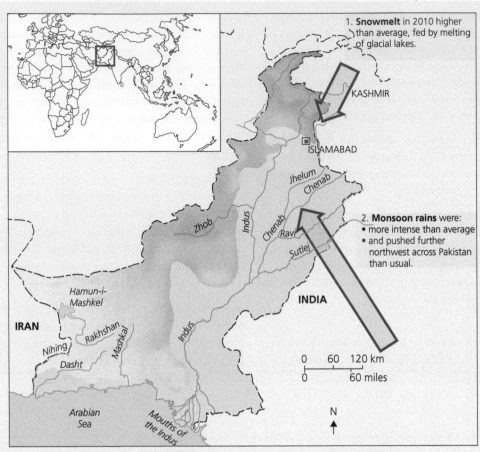

Figure 14 Map of the Indus, showing the causes of the 2010 flood

in 2010 was particularly intense, both in terms of the volume of water it deposited onto Pakistan and in terms of the areas of the country affected. Between 27 and 30 July, the northwest province of Khyber Pakhtunkhwa (KPK) was subjected to 60 hours of continuous rainfall totalling over 200 mm. In those four days alone, 130% of the normal 3-month rainfall totals fell, producing what the National Weather Forecasting Centre in Islamabad described as 'once in a century' levels of rainfall. As a result, a huge flood peak of discharge started to travel south down the Indus into central Pakistan. River flow peaked at 32,000 m³ s at the Sukkur Barrage. In fact, the floods devastated Sindh Province, but the province itself experienced comparatively little rainfall.

After these intense few days, the monsoon continued to feed water into the river, before two

further intense rainstorms in mid-August sent two further flood peaks south.

But it wasn't just the totals that were unusual; the monsoon rains extended further north than normal. Typically, northern KPK and the mountainous region of Kashmir experience only scattered rains during the monsoon season. This rain produced intense flash floods in the steep valleys, resulting in more deaths here than anywhere else in the country.

Human causes

The exceptional weather events of 2010 triggered the flooding. But its extent was magnified by several human factors.

Deforestation

Deforestation in Pakistan was a key factor in the flooding. Rates of deforestation in the country are around 2.4% per year. At this rate, the country's forest cover will be reduced to half of its 1995 level by 2020. Illegal logging compounds deforestation rates. It is thought that more than 70% of the forests in Malakand, KPK were cut down when the Taliban were in control of that region. With the trees removed, there was less interception of these rains, meaning that both the volume and speed of water reaching the rivers increased, contributing to the exceptionally high flood peak levels.

Extensive irrigation resulted in a loss of channel capacity

Since independence in 1947, the Pakistani authorities have engaged in an extensive programme of dam and canal building in order to provide irrigation for agriculture. It is estimated that 74% of the entire discharge of the river system is taken out along the course of the river and used for irrigation. But the Indus is a sediment-rich river, and this loss of discharge has meant that there is less energy to transport the sediment downstream to the floodplains or delta. Consequently, channel capacities have been significantly reduced, leading to these flood waters reaching flood levels more quickly.

Issues with levées

The extensive network of levées also added to this problem of loss of channel capacity. By cutting the river off from the floodplain with levées, sediment was confined to the channel and deposited on the riverbed, raising its level. Furthermore, there was levée failure during the flood event. For example, the Tori Bund levée in Sindh Province had lost 1.7m from its designed height due to erosion and poor maintenance prior to 2010 and so it was breached even before the flood peak arrived. Locals attempted to fix the breach by removing soil from the tops of other parts of the levée system, reducing their overall height as well. When the flood peak arrived, the levée failed catastrophically, causing the deaths of 97 people.

The effects of the 2010 flood on people, property and the land

People

The impacts of the 2010 flood on people were **largely negative** in the short term. In total, it is estimated that the floods killed 1,781 people. Most of these fatalities occurred in the early days and weeks of the flood in the KPK region, where very high flood waters were funnelled down steep mountain valleys. In addition, 2,966 people were injured and over 6 million people displaced from their homes. After the first week of flooding, 300,000 refugees had fled to Sukkur city in Sindh Province. The 20 relief camps quickly filled and people ended up sleeping rough wherever they could find shelter on the streets. The UN estimated that by mid-August, the flooding had in some way affected a staggering 14 million people. For example, 70% of the country's population, mostly in rural areas, did not have access to proper nutrition.

Over the longer term, however, the regular flood events that Pakistan experiences can bring some **benefits to people**. For example, the irrigation water from the river and the improvements in soil fertility resulting from more normal flood events than 2010 help boost agriculture in the country. The agricultural industry employs 45% of the country's population and contributes around 22% of its GDP. Without the flood waters of the Indus, there would be more people in poverty in the country.

Property

The impact of the 2010 flood on property was **very negative**. In total, 11,000 villages were flooded and 1.2 million homes were damaged or destroyed. The worst affected provinces were Punjab (483,000 homes) and Sindh (296,000 homes) — both of which have high population densities — and KPK in the northwest (259,000 homes), as it was closest to the location of the heaviest rain and is located within steep valley topography. In the Swat Valley in KPK, the steep-sided terrain and very high discharges had a devastating impact on property. It lost every one of its 20 bridges, an entire neighbourhood in the town of Madyan and its hospital. Only sand and stone were left behind where buildings once stood. The government estimated that over 7,000 schools and 400 health facilities were washed away. The total cost of the damage to property was around $7 billion, nearly one-fifth of Pakistan's GDP.

Land

In the short term, the impacts on the land were **very negative**. At their full extent, flood waters covered one-fifth of the country and the Indus in the Sindh Valley was over 24km wide, more than 25 times its usual width. Across the whole country around 7 million hectares of the most fertile land was washed away. In western Punjab, 570,000 hectares of cropland were destroyed. The deliberate breaching of levées to alleviate the flood risk in urban areas exacerbated the impact on the land. The cost of lost agricultural revenue was estimated to be $1 billion.

However, in the longer term, flood waters can bring **benefits to the land** in Pakistan. It is thought that salinisation badly affects around 25% of irrigated land, and this has resulted in 1.4 million hectares being abandoned. However, the fresh water from the floods goes some way to reducing salinity levels in the soil, increasing agricultural output in the long term. Furthermore, 60% of Punjab's farmers are reliant on groundwater to grow their crops.

In conclusion, the short-term impacts of large-scale floods are mostly negative. However, in the longer term, flooding brings benefits to the land and people.

Summary

- There are various methods of river channelisation, including re-sectioning, realignment and dredging.
- The main reasons for channelisation are: to increase river efficiency and capacity in order to reduce the impacts of flooding where channelisation is used; to improve land drainage in agricultural areas; and to improve navigation in a river.
- More modern approaches to river management favour sustainable and environmentally sensitive management. This is needed for various reasons including: the limitations of channelisation as a flood control strategy; to restore natural rivers and increase habitat diversity; the increased threat of flooding due to climate change.
- There are various soft engineering strategies that can be used to manage rivers sustainably including afforestation in the upper course, upstream washlands, river restoration schemes, urban land-use zoning and sustainable drainage in urban areas.
- Significant flood events in the Mississippi, such as that which occurred in 2011, are triggered by physical causes including snowmelt and heavier rain than usual. Once these physical factors trigger the heightened flood risk, human factors such as deforestation and the tendency for development on the floodplain increase the scale of the floods.

- The massive flooding of 2011 in the Mississippi Basin had mostly negative impacts on people and property in the shorter term, resulting in 30 deaths, displacement of people from their homes, economic losses, damage to buildings and destruction of crops. However, there were some benefits to people and the land, including rejuvenation of marshland and helping to address the impacts of the 2010 oil spill in the Gulf of Mexico.
- Significant flood events in the Indus, such as that which occurred in 2010, are triggered by physical causes including snowmelt in the Himalayan foothills and the annual monsoon rains (these were both higher than normal in 2010). Once these physical factors trigger the heightened flood risk, the scale of the floods is increased by human factors such as deforestation, loss of channel capacity resulting from extensive irrigation and the failure of levées.
- The massive flooding of 2010 in the Indus had significant negative short-term impacts on people, property and the land, resulting in 1,781 deaths, displacement of people from their homes, loss of buildings and villages, and the destruction of crops.
- However, there were some longer term benefits of the 2010 Indus flood, including increased soil fertility (which boosts agricultural output), groundwater recharging and reduction in the area of soils adversely affected by salinisation.

■ Topic 2 Ecosystems

Global biomes

An **ecosystem** is a community of plants and animals interacting with each other and the environment in which they live. **Biomes** are the large-scale global ecosystems that occupy distinct regions across the Earth. The main factor that determines the characteristics and location of these biomes is **climate** (temperature and precipitation).

Exam tip

Make sure you can refer to specific places to illustrate the location of global biomes.

The global distribution of biomes

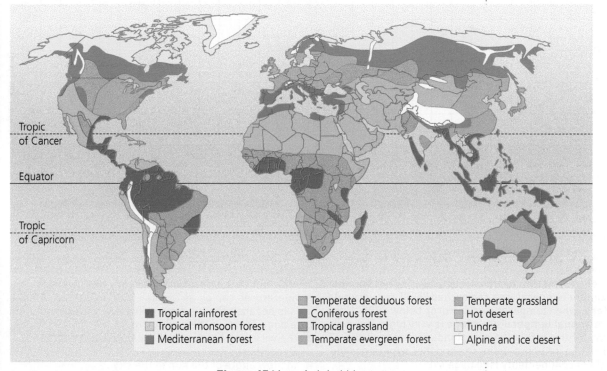

Tropic of Cancer

Equator

Tropic of Capricorn

■ Tropical rainforest	■ Temperate deciduous forest	■ Temperate grassland
▨ Tropical monsoon forest	■ Coniferous forest	■ Hot desert
■ Mediterranean forest	■ Tropical grassland	□ Tundra
	■ Temperate evergreen forest	□ Alpine and ice desert

Figure 15 Map of global biomes

Table 1 Global biomes

Tundra — found in the extreme high latitudes, generally polewards of around 60°. It includes northern Alaska and Canada, Greenland and northern Russia. Due to the extreme cold temperatures here, there are no trees and only low-growing hardy plants can survive.
Temperate grassland — found in the mid-latitudes between 30° and 40° N/S. The bands in the northern hemisphere are more extensive and are found in the continental interiors of North America (the Prairies) and southern Russia, and north Asian countries such as Mongolia. In the southern hemisphere, the temperate grasslands are less extensive. They are found on the eastern coast of South America (the Pampas), and further inland in South Africa (the Veld) and Australia (the Murray-Darling Basin).
Hot desert — located between 15° and 30° N/S where the easterly trade winds blow. They are mostly found towards the western sides of the continents. These include the deserts of Mexico and southwestern USA, the Atacama in South America, the Namib desert of southern Africa and the Australian desert. The Sahara desert of northern Africa extends from the western side of the continent right across to the east.
Tropical rainforest — found generally within 5° N/S of the Equator. They include the Amazon Basin in South America, the Congo Basin in west Central Africa, and the southeast of Asia.

The climate and soils of the tundra and temperate grassland ecosystems

Knowledge check 9

Compare and contrast the climate of the temperate grassland and tundra ecosystems.

The climate and soils interact to influence the vegetation found in biomes. We will explore these interactions in two biomes: temperate grassland and tundra.

Table 2 The ecosystems of the tundra and temperate grassland

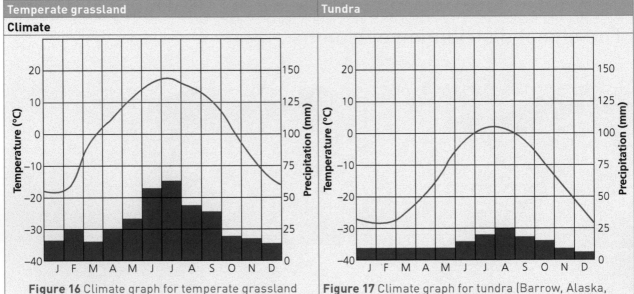

Temperate grassland	Tundra

Figure 16 Climate graph for temperate grassland (Saskatoon, Saskatchewan, Canada)

Figure 17 Climate graph for tundra (Barrow, Alaska, USA)

Temperature
Temperate grasslands are characterised by **hot summers** (maximum temperatures can exceed 35°C) and very **cold winters** (when temperatures can drop below −30°C), and so the **annual temperature range is very high**. The main reason for this is that the grasslands are far from the moderating influence of the sea and thus experience the effects of continentality (see page 49).

Precipitation
Overall, the levels of precipitation are **low**, with a typical range of 400-500mm per year in the North American prairies, as they are far from the sea. Furthermore, even when rainfall totals are relatively higher in the spring and summer months, the warm temperatures mean that there is **significant evapotranspiration**. This further reduces the amount of water available in the soil. The rain can fall in intense thunderstorms and therefore much of it flows to rivers via overland flow without entering the soil.

Temperature
Winter tundra temperatures are **very low**. Average temperatures can be as low as −30°C in Barrow, Alaska. This is because the tundra receives very little insolation (see page 44) from the sun due to:
- the very long hours of darkness (20 hours on 21 December)
- the low angle of the sun in the sky.
During the summer, despite the extremely long days (21 hours on 21 June) and limited cloud cover, temperatures can still struggle to reach even 10°C, and sometimes barely rise above freezing. This is largely due to the sun's very low angle in the sky.

Precipitation
Precipitation levels are **very low** as high pressure dominates the tundra areas. For example, the annual precipitation in Barrow is only 110mm. Most of this precipitation falls as snow.

Temperate grassland	Tundra

Soils

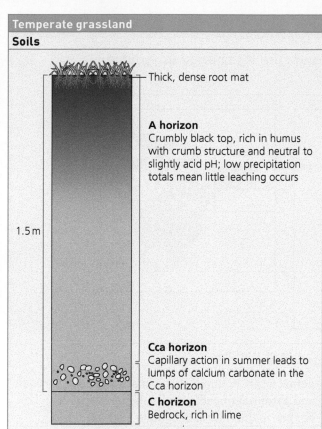

A horizon
Crumbly black top, rich in humus with crumb structure and neutral to slightly acid pH; low precipitation totals mean little leaching occurs

Thick, dense root mat

1.5 m

Cca horizon
Capillary action in summer leads to lumps of calcium carbonate in the Cca horizon

C horizon
Bedrock, rich in lime

Figure 18 Soil profile for mollisols

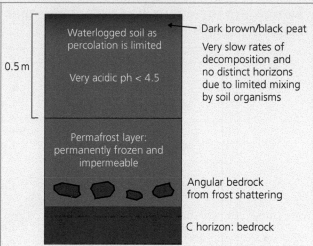

0.5 m

Waterlogged soil as percolation is limited

Very acidic ph < 4.5

Dark brown/black peat

Very slow rates of decomposition and no distinct horizons due to limited mixing by soil organisms

Permafrost layer: permanently frozen and impermeable

Angular bedrock from frost shattering

C horizon: bedrock

Figure 19 Soil profile for tundra

The soils found in this ecosystem are known as **mollisols** (or **chernozems**) and are deep (around 1.5 m in depth), rich soils consisting of two main layers or horizons (Figure 18). The **A horizon** is a crumbly, black topsoil that is rich in **humus**. During the summer, the soil organisms such as worms help decompose the litter and incorporate it into this top layer, forming the rich humus.

As precipitation only slightly exceeds levels of evapotranspiration the soils experience limited **leaching** (the removal of nutrients from upper levels of the soil by percolating rainwater), and so there is little loss of nutrients via this output. The leaching that does occur is mostly associated with the spring snowmelt and the heavy summer storms. In fact, the high summer temperatures and lack of soil moisture actually draw water upwards from the lower levels by **capillary action**. This brings nodules of calcium carbonate into the upper C horizon (known as the **Cca horizon**).

Mollisols have a crumb structure and a neutral to slightly acid pH.

The overall result of these soil characteristics is that mollisols are among the most naturally fertile soils in the world.

Given the very low temperatures in the tundra, at a depth of around 0.5 m there is a permanently frozen layer called the **permafrost**. It acts as an impermeable layer and makes it almost impossible for deep-rooted plants such as trees to grow. Above the permafrost, the soils:

- are waterlogged, as percolation is limited by the permafrost
- have very slow rates of decomposition of organic matter in the cold, wet conditions, as there are few soil organisms to aid decomposition
- have no distinct layers (horizons) again due to the limited mixing resulting from few soil organisms
- and are very acidic, with pH values that can reach below 4.5.

During summer, when temperatures may rise above freezing, there may be some limited leaching via water percolating down towards the permafrost. Frost-heave may move some rock fragments from the bedrock towards the surface.

The soils experience an annual cycle of freezing and thawing. In the highest latitudes, thawing can extend as deep as 30 cm. It extends further down into the soil in the lower latitude tundra areas. As a result, this active freeze/thaw layer of the soil experiences erosion and soil movement.

Due to the lack of plants able to fix nitrogen in the ecosystem, the soils have both limited nutrient levels and limited fertility.

Case study

Evaluating the actual and potential impacts of climate change on tundra ecosystems (regional scale): Alaska

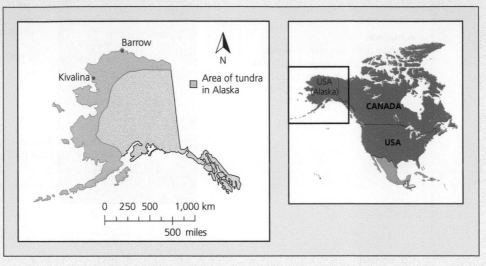

Figure 20 The location of the tundra areas in Alaska

In the past three decades, Alaska has warmed at twice the rate of the global average, so it makes an interesting ecosystem in which to examine the **actual** impacts (those that can already be observed) and **potential** impacts (those thought likely to happen in the future) of climate change.

Table 3 Actual and potential impacts of climate change in the tundra

Issue	Actual impacts	Potential impacts
Rising temperatures	Over the past 50 years, **temperatures** across Alaska increased by an average of 1.9°C, twice the national US average over that same period of time. But warming during the winter was even more dramatic, rising by an average of 3.5°C.	Average annual temperatures in Alaska are projected to increase an additional 2–4°C by the middle of this century.
	Evaluation	
	Rising temperatures have many negative impacts (see below), but there are also some **benefits** resulting from warmer summers. The number of growing days has increased by 20%, with benefits for agriculture and forest productivity on some sites. The impacts on aquaculture and fishing are mixed. Historically, modest increases in ocean temperatures have led to increases in salmon numbers. However, the complexities of the interactions within the ecosystem mean that it is uncertain how further warming could impact aquaculture, and negative impacts such as increased jellyfish numbers (which compete with fish for food and clog fishermen's nets) could outweigh any other benefits.	
Thawing permafrost and ice	**Actual impacts** Actual impacts The permafrost in Alaska has already begun to melt in response to the rising temperatures over the past 50 years. Soil temperatures at 1 m depth have risen by 4°C and at 20 m deep they have increased by 2.5°C.	**Potential impacts** Models project that permafrost in Alaska will continue to thaw and some models project that near-surface permafrost will be lost entirely from large parts of Alaska by the end of the century.

	Evaluation
	The impacts of this are **mostly negative**. More than 70% of the land in Alaska is vulnerable to subsidence if thawing occurs. As thawing continues into the future, it is estimated that the cost of maintaining infrastructure such as buildings, pipelines, roads and airports will increase by between $3 and $6 billion by 2030. Uneven sinking of the ground in response to permafrost thaw is projected to add between $3.6 and $6.1 billion (10–20%) to current costs of maintaining public infrastructure such as buildings, pipelines, roads and airports over the next 20 years. The melting of ice is already having negative impacts on transportation. Many of Alaska's highways are constructed on the permafrost — as it melts, these roads subside. Since the 1980s, the number of days when travel has been permitted on the permafrost has fallen from 200 to 100. The lack of snow in 2015 saved the city of Anchorage an estimated $1 million on snow removal that year. But in rural parts of Alaska, where there are few roads and people are dependent on travelling along frozen rivers, thin ice and open water meant that getting around was much more difficult that year. There are some **positive impacts** resulting from the melting of the permafrost. A reduction in permafrost will potentially make it easier to drill for oil in Alaska. The State gets 90% of its day-to-day finances from levies on oil and gas. However, in recent years, oil revenues have been in decline — at the same time that the cost of dealing with climate change has increased. For example, Alaska is currently having to deal with relocating climate change refugees; warmer temperatures are causing more coastal erosion in places such as Kivalina, an isolated community of 400 people in northwest Alaska, which will cost around $100 million to move. Extra oil revenues can help pay for this.

Lakes and wetlands	Actual impacts	Potential impacts
	The warming climate in the tundra has resulted in greater overall rates of evapotranspiration. Already, in the last 50 years in the southern two-thirds of Alaska, the area covered by lakes has decreased overall. Despite some localised increases in lake sizes due to permafrost melting, when we look at the State as a whole, lakes and wetlands are decreasing.	Future permafrost thaw will be likely to increase lake area in areas of continuous permafrost and decrease lake area in places where the permafrost zone is more fragmented.

	Evaluation
	The impact of lake and wetland change is **negative**. In the future, the continued drying of Alaskan lakes and wetlands could affect waterfowl management in the USA. Alaska has 81% of the National Wildlife Refuge System and provides breeding habitat for millions of migratory birds that winter in more southerly regions of North America and on other continents. The changes in lakes in this ecosystem have the potential to disrupt the habitats for these migratory birds. In addition, waterfowl are an important food source for Alaska Natives and other rural residents. Loss of wetland would reduce the waterfowl numbers, affecting the food chain and reducing food supply to local people.

Increased wildfires	Actual impacts	Potential impacts
	Climate change is producing warmer, drier summers with increased frequency of thunderstorms. This — along with loss of wetland — has resulted in an increase in wildfires in the tundra. There were more large fires in Alaska in the 2000s than in any decade since record-keeping began in the 1940s.	Despite the fact that projections suggest that precipitation levels could increase by 15–30% by the end of the 21 century, nevertheless increased evaporation associated with rising temperatures is expected to reduce water availability overall in the biome. This is expected to increase the fire risk. Even if climate warming was curtailed by reducing greenhouse gas emissions, the annual area burned in Alaska is still projected to double by mid-21st century and to triple by the end of the century.

	Evaluation
	The impacts of burning are **partly positive** and **partly negative**.
	The effects of the wildfire on habitats are mixed. On one hand, the burning improves habitats for plants such as berries and mushrooms, and creates conditions more favourable for moose. However, the habitat for the caribou is badly affected. The lichen on which they depend in the winter months take between 50 and 100 years to recover following burning. Caribou in turn are a critical food source for predators such as bears and wolves, as well as for Alaska Natives.
	Furthermore, exotic plant species that were introduced along roadways are now spreading onto river floodplains and recently burned forests, potentially changing the suitability of these lands for timber production and wildlife. Some invasive species are toxic to moose, on which local people depend for food.
	In addition, the fires have a negative impact on the ability of the tundra to store carbon dioxide. A single large fire in Alaska in 2007 released as much carbon into the atmosphere as had been absorbed by the entire Arctic tundra during the previous quarter-century.
Overall evaluation	Overall, despite some benefits in terms of increased growing season and funding resulting from oil exploitation, the impacts of climate change on the tundra ecosystem are largely negative.

Exam tip

In any evaluation question, you should always finish with a short concluding sentence giving your overall evaluation of the issue.

Exam tip

Make sure you can refer clearly to both actual and potential impacts of climate change on the tundra.

Summary

- The locations in which global biomes (including tundra, tropical rainforest, hot desert and temperate grassland) are found are largely determined by climate.
- Temperate grassland biomes are characterised by hot summers and cold winters. Precipitation levels are low overall and there is significant evapotranspiration during the heat of the summer months.
- The soils found in temperate grasslands are called mollisols and consist of a well mixed A horizon sitting over a C horizon. They experience limited leaching and some capillary action during the summer. They are very fertile soils.
- Tundra biome temperatures are very low in the winter and quite low during the summer. Precipitation levels overall are very low.

- The low temperatures of the tundra regions mean that the soils here are underlain with a permafrost layer. The soil has no distinct horizons due to limited work by soil organisms in the cold conditions. The soils have limited fertility.
- Climate change in the tundra is already having a number of largely negative effects. The thawing of permafrost is damaging infrastructure. Warming temperatures are increasing the number of damaging wild fires and reducing the amount of lake and wetland coverage. These impacts are projected to increase throughout the 21st century.

Small-scale ecosystems

Biotic and abiotic components

Biotic components — the living organisms (plants and animals) of the ecosystem. The total amount of living organic matter in an ecosystem is known as the **biomass**.

Abiotic components — these are the non-living components of the ecosystem, including **soils** and **climate**.

The ecosystem as a system

We can use the **systems** approach to analyse ecosystems. Look out for references to inputs, outputs, stores and transfers in the following subsections.

Table 4 Ecosystem as an open system

	Inputs	Stores	Transfers	Outputs
Energy	Photosynthesis	Flora and fauna (stored as sugars and proteins)	When organic matter is consumed by heterotrophs	Respiration
Nutrients	Precipitation and weathering	Biomass Litter Soil	Fallout Decay Uptake	Run-off and leaching

Energy flows and trophic structures

The key concept here is that **energy flows through the ecosystem**. This means that it **enters, moves through** a series of trophic levels and then **leaves** via heat. Figure 21 outlines how this process occurs and introduces you to a range of terms that are used to describe elements of the process.

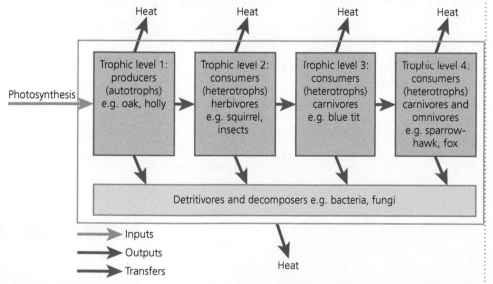

Figure 21 Energy flow diagram, illustrated with reference to Breen Wood, Ballycastle, Co. Antrim

Exam tip

There are many terms in this topic, many of which will be new to you. Remember that top-level answers require *extensive use of geographical terms*, so you should learn them all and be able to use them appropriately in your answers.

Inputs of energy

Energy enters the ecosystem via photosynthesis, as green plants fix solar energy and convert it into chemical food energy.

Food chains and trophic levels

This input of food energy is the basis for a **food chain**, where the energy fixed by photosynthesis becomes available to subsequent levels of the ecosystem. Each level is known as a **trophic** (energy) **level**.

The first trophic level consists of the plants. As they can produce their own food directly, they are called **producers** (or **autotrophs**). All subsequent trophic levels have to consume plants and other organisms in the previous level, so they are called **consumers** (or **heterotrophs**).

In the second trophic level, the **herbivores** (plant eaters) consume plants from the first level. In turn, these herbivores provide food for the third trophic level, the smaller **carnivores** (meat eaters). The fourth, and usually final, trophic level is made up of the larger carnivores and **omnivores** (plant and meat eaters).

Thus food energy is transferred up through the trophic levels. However, on average only around 10% of the energy is transferred up to the next trophic level. The reasons for this are as follows.

- Energy is lost via life processes in organisms such as **respiration** (which ultimately causes energy to be outputted from the ecosystem as **heat**).
- The transfer of food energy is **incomplete** — not everything that can be eaten is actually eaten by animals in the next trophic level (e.g. animals might die of old age rather than being consumed by another animal).
- The transfer of food energy is **inefficient** — not everything actually eaten can be metabolised by animals in the next trophic level. This results in excrement (which is broken down by the **detritivores** and leaves the ecosystem, rather than being passed up the trophic levels).

Due to this energy loss, there are usually **no more than four trophic levels** in an ecosystem.

At each trophic level, the detritivores (such as bacteria, maggots and fungi) are consumers that act to decompose dead organic matter.

The energy available at different trophic levels can be shown in a trophic pyramid (Figure 22).

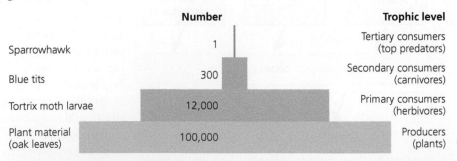

Figure 22 Trophic pyramid for the deciduous forest at Breen Wood, Ballycastle, Co. Antrim

Knowledge check 10

What are the terms you could use to describe the organisms at trophic levels 2 and 3?

Nutrient cycles

The key point here is that **nutrients mainly cycle around between stores within the ecosystem** (although there are inputs and outputs as well). This is shown in Figure 23 — note the inputs, outputs, stores and transfers.

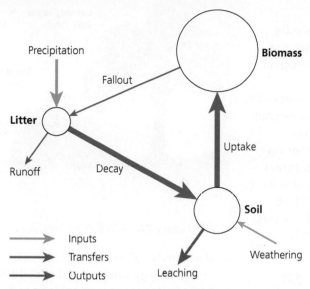

Figure 23 The Gersmehl model applied to Breen Wood (deciduous forest)

The nutrients leave the **biomass store** and are transferred by **fallout** into the **litter store**. This dead organic matter is broken down by the **decomposers** and transferred by **decay** into the **soil store**. Here, the nutrients are available to plants via their roots and are transferred by **uptake** into the **biomass store** and thus the cycle continues. In addition, there are two **inputs** of nutrients (such as carbon dissolved in **precipitation** or minerals **weathered** from the bedrock) and two **outputs** (losses via **runoff** from the **litter store** or **leaching** from the **soil store**).

The diagram can be drawn proportionally to show the relative amounts of nutrients in the stores and transfers: the larger the circles, the greater amount in the stores; the thicker the lines, the greater the volume being transferred.

The **rate of nutrient transfer** can be affected by factors such as:
- **climate** (in hot, wet tropical rainforests the rate of decomposition is very high and rates of leaching are also considerable)
- **soils** (acidic soils have fewer soil organisms and thus decomposition rates are slower)
- **vegetation type** (the leaves from coniferous trees take longer to decompose than those from deciduous trees).

Case study

Deciduous forest in Breen Wood, Ballycastle, Co. Antrim

Abiotic components of Breen Wood deciduous woodland

Breen Wood deciduous forest near Ballycastle is one of the few remaining areas of the ancient deciduous woodlands that used to cover much of Ireland. Its **abiotic** components are:

- **climate** — average temperatures range from 4°C in winter to 15°C in summer and annual rainfall totals are quite high at around 1,600 mm.
- **soils** — the **podsols** found here are generally poor in quality, partly due to the basalt parent rock, which is low in nutrients. Thus, the soils are thin and have an acidic pH of around 4.5. Furthermore, the combined high rainfall totals and sloping relief ensure that **leaching** is significant.

As a result of these abiotic factors, the range of plants and animals is smaller than would typically be expected in a deciduous forest and the 200-year-old oak trees in Breen Wood are about half the size they would be in lower areas.

Trophic structure in Breen Wood deciduous woodland (biotic components)

The first trophic level consists of the **producers** in Breen Wood. Energy is inputted from the sun as these **autotrophs** are able to fix solar energy via **photosynthesis**. There are three main layers of vegetation in Breen Wood. The first is oak and birch trees, forming a dense **canopy layer** 20 m above the ground. Below these, a **shrub layer** is found consisting of hazel and holly trees. In the **ground layer**, plants such as ferns, brambles and mosses grow in the damp shade below the trees (Figure 24).

Only about 10% of the energy fixed by the autotrophs is available to the subsequent trophic levels. There are various reasons for this including:

- the process of respiration uses up energy which is then outputted from the ecosystem as heat
- the transfer of energy is incomplete and not everything that can be eaten is actually eaten by animals in the next trophic level

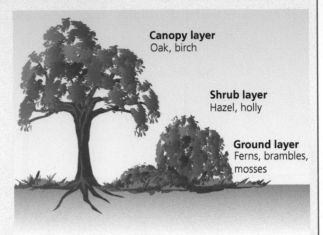

Figure 24 Layers of the deciduous forest

- the transfer of food energy is inefficient and not all energy can be metabolised efficiently so some is lost via excrement (and removed by the detritivores rather than being passed up the trophic levels).

The 10% of the energy fixed by the first trophic level is available to the **primary consumers** in next trophic level, as the **herbivores** eat the vegetation. The herbivores in Breen Wood are mostly insects, including around 15 species of butterfly (orange tip and speckled wood), and red squirrels. The harsh climatic and soil conditions mean that the range of **heterotrophs** in the ecosystem is more limited than in other deciduous woodland areas.

Again, about 10% of the energy from trophic level 2 is available to the **secondary consumers** in the third trophic level, as they consume animals from the second trophic level. This level consists of birds, including **omnivores** such as blue tits and goldcrests, which forage for seeds and insects.

Any remaining energy (about 0.1% of the energy fixed at trophic level 1) is passed on to the fourth and final trophic level. This consists of **tertiary consumers**: **carnivores** and **omnivores** including sparrow hawks and buzzards (which feed on the smaller birds), badgers (which feed on earthworms, small birds, fruit and nuts), foxes (which eat the squirrels and smaller birds, as well as grass and fruit) and stoats (carnivores that feed on squirrels and birds).

In addition, at each trophic level, detritivores such as earthworms, fungi and maggots help decompose dead organic matter. In Breen Wood, due to the acidic podsol soils, the work of the detritivores is limited and decomposition rates are slow.

Nutrient cycles in Breen Wood deciduous woodland (biotic components)

(Refer to Figure 23, p. 37 as you read this.)

There are three main stores of nutrients in Breen Wood. The largest by far is the **biomass store**, consisting of the plants (such as oak, holly and hazel trees, ferns and mosses) and the animals (including orange tip and speckled butterflies, red squirrels, hawks, foxes and badgers). During the autumn, the deciduous trees shed their leaves to reduce moisture loss via transpiration during the winter, contributing fresh leaf litter via the **fallout pathway** to the next store: the **litter store**.

The litter store is smaller than the biomass store, but is larger than typical for a deciduous forest due to the acidic podsol soils and cool average summer temperatures of 15°C in Breen Wood. This means that there are fewer soil organisms such as earthworms and so the rates of decomposition are quite slow — it may take years for the the detritivores, such as earthworms, to full break down the leaves and for them to travel via the **decay pathway** to the soil store.

The **soil store** is of similar size to the litter store. The **uptake** of nutrients by plants in a deciduous forest is only about 25% as efficient as a tropical rainforest, so the soil store remains quite large. However, the outputs of nutrients are slightly higher than a typical deciduous forest — the high rainfall totals (around 1,600 mm) and cool summer temperatures (averaging around 15°C), along with the sloping relief of the area, mean than leaching and **runoff** are quite high, causing nutrients to be lost from the system. In addition, the inputs of nutrients are lower than usual — the basalt bedrock does not release many nutrients via **weathering**, leaving the main nutrient input to come from **precipitation**.

Exam tip

Facts are always essential in case studies. This normally means figures and places. In this case, however, your facts are mostly plant and animal names. Make sure you learn them and include them in exam answers.

Plant succession

Plant succession refers to the processes of change in the plants — along with the soils and microclimate — in an ecosystem (Figure 25). During succession, plant species replace one another until a climax is reached that is in balance with the environment.

Plant succession A series of changes that take place in vegetation over time as one main plant species replaces another. Primary succession is succession that occurs on previously unvegetated surfaces, for example sand dunes or bare rock. Secondary succession occurs on previously vegetated surfaces that have lost their vegetation through processes such as fire or landslides.

Microclimate Small-scale climatic characteristics of an area that differ from the general climate. It can apply to areas ranging from a few square metres to many square kilometres.

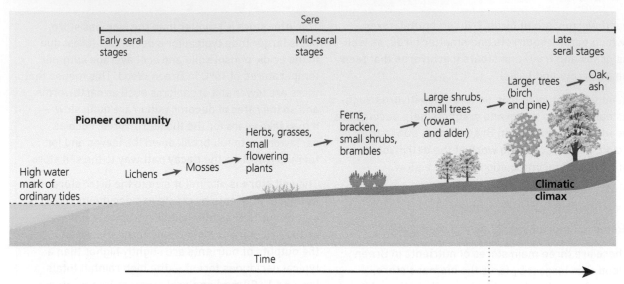

Figure 25 An example of plant succession

To understand how succession occurs, it is important to realise that various **processes of environmental change** occur to produce it. These obviously include changes in vegetation, but vital changes also occur in soil and microclimate. The following key points help to explain this.

- The plants **most adapted** to the particular environmental conditions of soil and microclimate **dominate** at that given time (e.g. mosses colonise bare rock).
- However, these very plants tend to **modify the environment** in which they grow, improving the **soil conditions** by adding organic matter, increasing soil depth, aiding soil moisture retention and changing soil pH. They also improve the **microclimate** by reducing ground wind speed and sheltering the soil, thus reducing evaporation and maintaining higher soil temperatures.
- Thus they create **conditions more favourable for other plant types**, which will then come in and replace the previous plants (e.g. the moss will be replaced with taller plants such as grasses, which will in turn be replaced by shrubs). In addition, as the soil and microclimate become less harsh, a wider variety of plants can grow there, so the number of plant species tends to increase too, although each seral stage tends to have one dominant plant type.

This means that change is self-perpetuating, as plants invade, dominate, change the environment and are replaced, until a climatic climax is reached and the plants reach a balance with the environment. Once the climatic climax vegetation is established, it tends to exclude its rivals and the number of plant species can drop slightly.

Exam tip

It should be clear from these points that while the plants certainly change during succession, there are also changes to the soils and the microclimate. Read the exam questions carefully to see which of these three you are being asked to discuss.

Seral stage The name of each stage in the process of plant succession; the whole sequence of change is called a **sere**.

Climatic climax The final stage of plant succession, when the vegetation has reached a balance with its environment.

Knowledge check 12

State the three aspects of the environment in an ecosystem that can change as plant succession occurs.

Table 5 Changes during succession

	Explanations for changes
Soil changes	
Depth increases	Due to the addition of more organic matter from the biomass and more weathering of the bedrock
Humus increases	More organic matter is added to the soil via nutrient cycling
Moisture increases	The humus aids moisture retention and the shade from plants reduces evaporation from the soil
Stability increases	The increasing root system binds the soil together
pH decreases	The soil becomes less alkaline due to the addition of humic acids
Colour darkens	More organic matter is added to the soil via nutrient cycling
Plant changes	
Biomass and plant longevity increase	The deeper and richer soil can support larger and more long-lived plants such as trees. Plant abundance increases and the amount of land surface not covered by plants decreases
Species diversity increases	The improving soil and microclimatic conditions can support a wider variety of plant types, not just those adapted to the harsh conditions of colonisation
Stratification (different layers of vegetation)	The wider variety of plant types increases stratification
Microclimate changes	
Wind speed drops	The increased plant cover provides shelter and acts as a wind break

Colonisation The process whereby the initial plants invade and begin to grow on bare rock or soil; these initial colonising plants are called the pioneer species.

Plagioclimax

Plagioclimax refers to the vegetation community that is found when human activities interfere with the natural sequence of succession and prevent it from reaching its climatic climax. This can result from activities such as deforestation, burning, draining or grazing.

One example of a plagioclimax is the heather moorlands that are found in many upland areas in the British Isles. The climatic climax here should be deciduous forest. However, these forests have been cleared and the areas used for sheep grazing, allowing the heather plant to dominate. The ecosystem is held at plagioclimax in part by the sheep grazing (which cuts back young tree saplings), but the heather is also burned on a 15-year cycle (a management strategy known as **muirburn**) to return nutrients to the soil and encourage new heather growth. This also prevents succession carrying on towards climatic climax.

> **Exam tip**
>
> Include an example of a cause of plagioclimax (e.g. muirburn in heather moorlands) in your exam answers, expanding into some more detailed discussion of how one cause operates if there are more than 2 marks available for the answer.

A psammosere succession in the sand dunes at Portstewart Strand

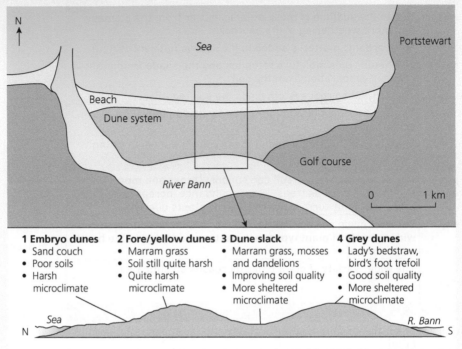

Figure 26 Succession in the psammosere at Portstewart Strand

The initial bare surface on which colonisation occurs is the sand at the very top of the beach. The conditions here are very tough: the soil has next to no organic matter, the salt levels are high, the pH is very alkaline due to the amount of sea shells in the sand and the microclimate is harsh, with no shelter from strong onshore winds blowing in off the Atlantic Ocean. Therefore, only very hardy plants adapted to these conditions may grow here. In Portstewart Strand this is mainly sand couch. This initial seral stage is referred to as **embryo dunes**.

As these plants colonise, they begin to change the environment, especially the soil. Their roots help to stabilise the soil and their leaves help the dunes to grow in height as they trap more sand. The soil slowly begins to improve as more organic matter is added when plants die. This begins to increase the humus levels, aiding water retention in the very sandy soils.

These improving conditions create the conditions more suited to other plants, especially marram grass. Marram grass begins to invade and dominate, forming the next seral stage — the **foredunes** or **yellow dunes**. Marram grass is particularly adapted to the still quite harsh conditions in the yellow dunes. Its leaves are tightly rolled and have a shiny surface to reduce moisture loss via transpiration. Its growth is stimulated by burial under sand and spreads readily via its rhizome root system. These rhizomes further stabilise the soil and increase the dune height.

The sand dunes at Portstewart form a series of ridges roughly parallel with the beach. As you cross over the first ridge into the **dune slack**, the microclimate changes and becomes noticeably more sheltered. This, along with the improving soil quality, allows a wider variety of plant species to grow. At first, mosses grow in the spaces between the marram grass and rosette plants, such as dandelions, start to appear. These plants tend to spread laterally, covering the surface of the soil, reducing moisture loss via evaporation and helping the soil temperature to remain a little higher.

Further into the dune slack, these changing conditions allow plants such as ribwort to grow, and slowly the marram grass is replaced.

At the back of Portstewart Strand are the old, stable **grey dunes**. The pH here is between 6.5 and 7.5, and the soil has significant amounts of humus, supporting plants such as lady's bedstraw and bird's foot trefoil.

As this dune system is managed by the National Trust, the succession is prevented from reaching its climatic climax by strategies such as allowing cattle grazing, especially towards the eastern end, although the odd isolated willow or hazel tree can be found towards the back of the dunes.

Exam tip

As in the previous case study, the facts for Portstewart Strand are mostly plant types. Make sure you know and include a wide range of plants to represent the various seral stages in the dune system.

Exam tip

In longer exam questions on this case study you will probably be asked to describe *and* explain the successional changes. Make sure you are able to clearly explain the processes of change in soils and microclimate brought about by changing vegetation.

Knowledge check 13

Making reference to your small-scale study of plant succession, describe two of the above changes.

Summary

- An ecosystem is an open system consisting of abiotic components (climate and soils) and biotic components (plants and animals) interacting with each other.
- After entering the ecosystem via photosynthesis, energy flows up a food chain/web through a series of trophic levels. The energy transfers are very inefficient, however, and much of the energy is lost before it reaches the next trophic level. As a result, ecosystems seldom have more than four trophic levels.
- Nutrients cycle around between three stores within the ecosystem (biomass, litter, soil) through a series of transfers or pathways (fallout, decay and uptake). In addition, there are inputs (via precipitation and weathering) and outputs (via runoff and leaching) of nutrients from the ecosystem.

- Plant succession refers to the process of change in vegetation over time whereby one dominant plant species replaces another dominant species, through a series of seral stages, until the plants reach a balance with the environment known as the climatic climax.
- The initial pioneer plants must be adapted to harsh conditions of soils and microclimate. But they begin the process of improving the soil and microclimate, making the conditions more favourable for other plants. Thus vegetation coverage, plant height and species diversity increase as succession continues.
- If humans stop the natural process of succession by activities such as grazing or burning, the stage at which it is held is known as a plagioclimax.

■ Topic 3 Atmosphere

The processes that shape our weather and climate

The global energy balance

The Earth's atmosphere extends about 1,000 km from the surface of the Earth but the majority of the gases needed to support life are found within the lower 40 km of the atmosphere. Most of the weather and climate within the atmosphere takes place within the **troposphere** (which is about 16–17 km deep).

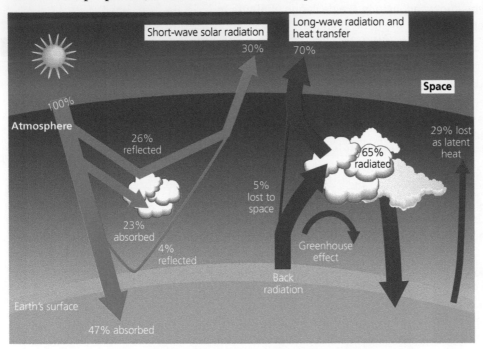

Figure 27 Global energy balance

The global energy balance is an open system where the Earth receives heat/solar energy (insolation), and energy is stored and transferred continuously between the Earth and its atmosphere. Some energy will eventually be reflected back into space. The amount of energy reflected depends on the Earth's albedo.

How is heat transferred around the globe?

Vertical heat transfer

Heat can be transferred from the surface of the Earth to the atmosphere, having a cooling effect on the Earth while warming the air. This can happen in one of three ways.

■ **Radiation:** the land radiates heat back out to space through the use of long-wave radiation.

> **Exam tip**
>
> Make sure that you understand the key terms and their definitions.
>
> **Insolation** Radiation/ energy that comes from the sun and enters the Earth's atmosphere.
>
> **Albedo** The amount of reflection that occurs when solar energy meets clouds, dust or the surface of the Earth. It is shown as a percentage of all of the incoming radiation.

- **Convection:** warm air is forced to rise as part of convection currents. The rising warm air is replaced by colder, descending air.
- **Conduction:** energy is transferred through contact.

Horizontal heat transfer

Heat across the globe is not distributed equally. Between the Equator and 40° N and S of the Equator there is a heat surplus, while from the poles to 40° N and S there is a heat deficit. The atmosphere redresses this in the following ways.

- **Ocean currents:** the North Atlantic Drift, for example, brings warm water from the Caribbean towards Europe while the Labrador current brings cold water down from the Arctic to the east of Canada (especially in the winter).
- **Winds:** winds carry surplus energy from the tropic areas away from the Equator (for example the tropical maritime air mass moving from the tropics towards the British Isles). As the warm wind blows over the sea (or vegetation) it picks up moisture and in the process stores latent energy, for example, the Arctic air mass moving from the north towards the British Isles).
- **Hurricanes:** a major weather system can sometimes occur that involves ocean currents, warm seas and tropical winds, which lock latent heat in and then transfer it gradually across the globe.

The role of ocean currents

Ocean currents play an important role in moving heat energy around the world. Insolation deeply penetrates the ocean waters. Warm water flows north and south from the Equatorial seas, warming colder areas. Water flows towards the poles because warm water is less dense than cold water. Cold waters move in to replace this water.

Knowledge check 14

Describe the three processes that can transfer heat vertically in the atmosphere.

Latent heat As water changes state from liquid to gas (evaporation), some energy is stored due to the process.

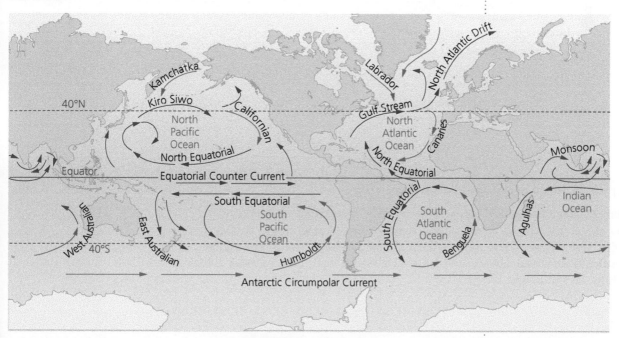

Figure 28 Ocean currents

The general circulation of the atmosphere

The general circulation of the Earth's atmosphere helps to explain the pattern of wind and pressure belts. This circulation system is the main way in which the Earth can **redistribute energy** received at the Equator towards the poles.

Exam tip

Make sure that you can describe and explain the way in which heat is transferred horizontally, making special reference to named ocean currents.

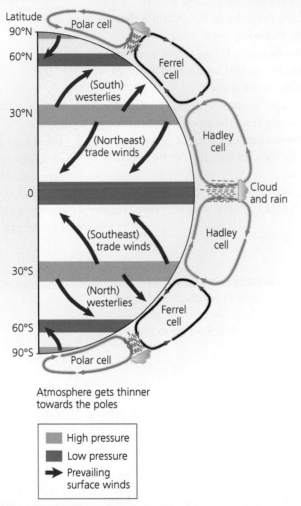

Figure 29 The general circulation of the atmosphere

Surface pressure belts and wind

Figure 29 shows the general pattern of pressure belts and wind within the atmosphere. It is a highly complicated system but can be simplified so that we can notice that winds usually move from areas of high pressure towards areas of low pressure. This starts a movement of air at surface level that can be opposed at higher levels in the atmosphere system.

The tri-cellular model

Three cells control the circulation of air (Figures 30–32):

- **Development of the Hadley cell**

  The Earth's atmospheric engine gets its energy from the direct solar heating at the Equator. Heat is transferred to the air above and the air rises and cools. Rising air means that low pressure, known as the inter-tropical convergence zone (ITCZ), is left at the surface.

  The rising air diverges and flows towards the poles (both north and south) and sinks again around 30° N and S of the Equator. This warms and produces high pressure. The air is then diverted back towards the Equator at a much lower level while some air moves towards the mid-latitudes to redress the global heat imbalance (Figure 30).

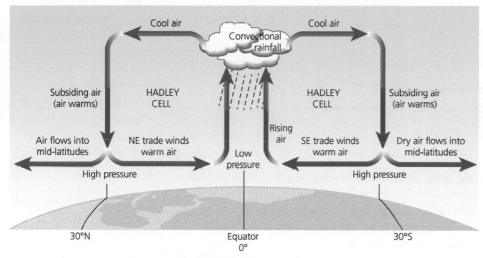

Figure 30 The Hadley cell

- **Development of the Polar cell**

  The extreme cold air at the poles leads to subsiding air (and high pressure at the surface). The air then moves away from the high pressures across the surface towards the 60° N (and S) point where it meets the warmer tropical air at what is called the polar front (Figure 31).

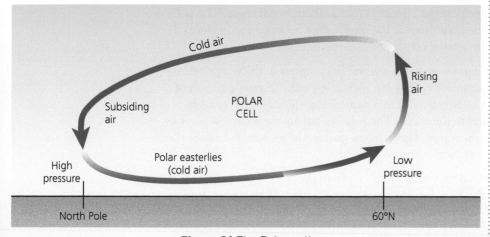

Figure 31 The Polar cell

■ **Development of the Ferrel cell**

Some of the air from the Hadley cell at 30° N (and S) moves across the surface towards the mid-latitudes (the westerlies). At 60° N (and S), this warmer air meets colder air from the poles. The polar front forms between them and the warmer air rises, leaving low pressure at the surface (Figure 32).

Knowledge check 15

Describe how heat energy is circulated from the Equator to the poles using the tri-cellular model.

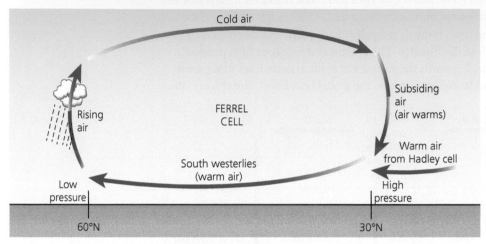

Figure 32 The Ferrel cell

Exam tip

You need to know the different air movements associated with each part of the tri-cellular model at both the upper atmosphere and surface levels. Practise drawing the diagram for each cell.

Moving from the Equator to the poles there are alternating bands of high and low pressure. However, in some cases other factors, including continentality, modify this pattern. For example, Siberia should be located in a band of low pressure throughout the year, but it experiences high pressure in the winter as a result of the extreme cold of the continental interior causing the air to subside.

Jet streams and upper westerlies

Increasingly, weather presenters refer to the impact that the movement of a jet stream can have on weather. A jet stream is a ribbon current of very strong wind that will move weather systems around the world. The air movement usually occurs between 7 and 15 km above sea level in the upper atmosphere, where winds usually move from the west to the east and can reach in excess of 200 mph.

The strongest jet stream is the polar jet stream in the northern hemisphere. The polar jet stream usually forms at the polar front at the intersection between the Ferrel and Polar cell due to the rising air around 60° N and S of the Equator.

The position of a jet stream can move due to natural fluctuations in the temperature between tropical and polar air masses. This can then cause either a steady stream of depressions to move across the British Isles in a sequence or the stream to entirely miss the British Isles and land to the south. The polar jet stream meanders its ways in a series of curves called Rossby waves. These waves start as the jet stream crosses the Rocky Mountains.

The upper westerlies are prevailing winds that move from the west to the east between 30° and 60° from the Equator. They move from an area of high pressure to an area of low pressure and form part of the Ferrel cell.

The factors that influence temperature

Latitude is the most important factor when looking at the pattern of global temperature. Areas at the Equator usually receive 12 hours of sunlight. The angle of the sun concentrates any rays at the Equator, but the angle becomes more acute towards the poles, with less energy being spread over a much larger area. However, variations are obvious within this pattern as other, more localised factors have their influence:

- **Continentality:** Areas that are much further inland experience bigger temperature ranges and extremes than areas on the coast. For example, the UK average winter temperature is between 0°C and 8°C, whereas Siberia on a similar latitude has average temperatures of between −20°C and −10°C.

- **Altitude:** As air rises, it cools, and so upland areas are generally cooler than lowland areas. The reason for this is because the air experiences a drop in pressure as it rises. This allows the air to expand — this expansion uses some energy and so the air temperature drops. This process is known as **adiabatic cooling**. On average, air cools adiabatically by around 0.6°C for every 100 m of height.

- **Seasonality:** The combined impact of these factors can cause a seasonal difference. There is a huge disparity in the average temperatures between summer and winter months. Summer in the northern hemisphere often means that maritime areas are cooler than inland areas. In the winter, the sea retains its heat for longer and therefore maritime areas can be warmer. Ocean currents, like the extension of the Gulf Stream as the North Atlantic Drift, also warm coastal areas further. In the winter, continental areas will lose/radiate more heat, which means that temperatures will fall below places with a similar latitude.

Knowledge check 16

Discuss how continentality might influence the pattern of temperature in an area.

The global pattern of precipitation

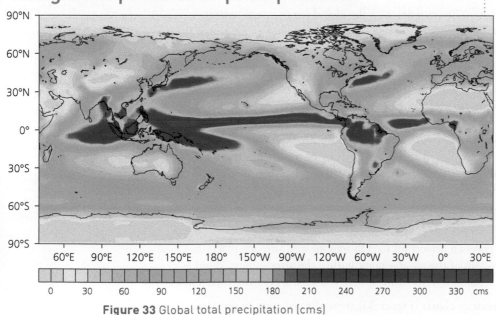

Figure 33 Global total precipitation (cms)

There is a close relationship between the global ocean currents and patterns of precipitation. The highest amounts of precipitation take place across the equatorial regions but other high areas of precipitation follow areas where warm flows of seawater are moving north and south away from the Equator (due to the Gulf Stream, the North Atlantic Drift, the North Pacific Current and the North and South Equatorial Currents). Higher pressure areas such as the Polar regions and the Tropics are noted for their low rainfall and desert/arid conditions.

Exam tip

Global patterns questions will often ask you to describe specific patterns, so make sure that you have practised doing this for precipitation, surface temperature, pressure and wind. Make sure that you use geographical locations in your answer and that you quote figures from the resource. Look for the high points, low points and any averages that might appear.

The global pattern of surface temperature

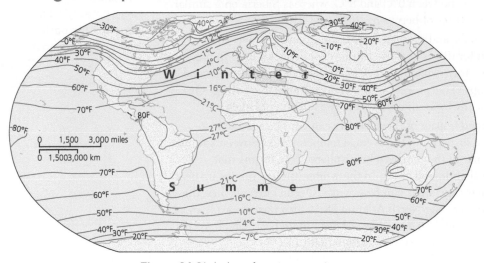

Figure 34 Global surface temperature

Any transfer of heat will have a direct impact on the global temperature pattern. The pattern across the world is extremely complex and is controlled by a number of important factors. The pattern is significantly different through the year — with temperature anomalies most obvious between the summer and the winter.

Figure 34 shows one representation of this map when it is winter in the north and summer in the south. The highest temperatures (27°C) are usually found around the equatorial region (through South America and Central to South Africa). The temperature generally decreases with lines of latitude (towards the poles) but some anomalies are noted in Western Europe (due to influence of continentality). One massive anomaly in the northern hemisphere is the particularly cold area in northern Russia (−34°C/−30°F and lower).

Exam tip

Make sure that you can use geographical terminology to describe the pattern of global temperature (use directions/continents/ lines of latitude as points of reference).

The global pattern of pressure

Atmospheric pressure plays an important part in moving air from one place to another. Pressure is the weight of the air at a point within the atmosphere. Wind is the movement of air that results from the atmosphere redressing pressure differences.

The two global surface pressure charts (Figure 35) show the variation in pressure between January and July. The arrows help to show the changes to wind directions through the summer and winter months.

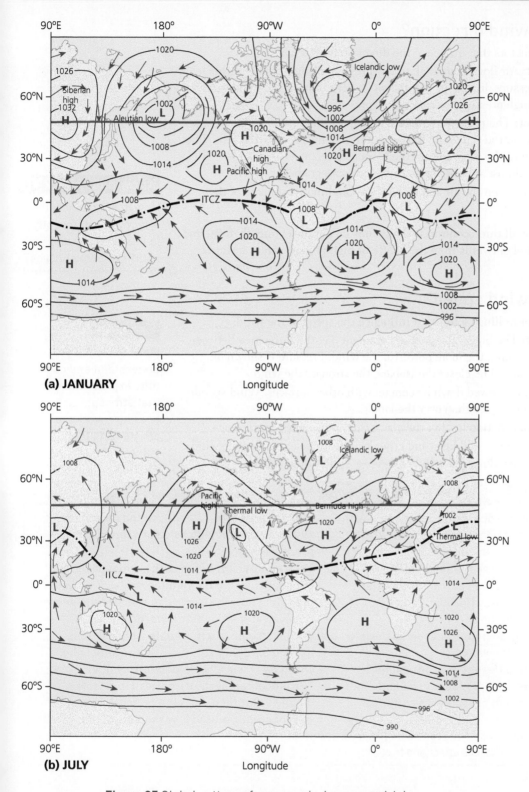

Figure 35 Global pattern of pressure in January and July

What controls wind direction?

There are three main factors that influence the direction in which wind blows:

- **Pressure gradient:** Air flows from high to low pressure along a gradient (the difference in the pressure divided by the distance). For example, the trade winds blow from areas of high pressure to areas of low pressure.
- **The Coriolis effect:** The rotation of the Earth deflects the flow of air in the northern hemisphere to the right (and in the southern hemisphere to the left). The further north a place is, the more impact the Coriolis effect has — areas 50–55°N have winds that are almost at right angles to the pressure gradient.
- **Friction:** Friction between the air and the surface can limit the impact of the Coriolis effect.

The resultant impact of all three controls is that winds do not travel directly from high to low pressure but spiral outwards from the centre of high-pressure areas towards areas of low pressure.

What controls wind speed?

There are four main controlling factors that influence the speed of the wind:

- **Pressure gradient:** The bigger the pressure gradient (i.e. the difference between the high and low pressure), the more powerful the wind. Pressure is shown on weather maps as isobars. The closer the isobars, the stronger the wind.
- **Friction:** Wind can be slowed down by contact with other surfaces. Wind speeds tend to be higher over the sea than over the land.
- **Turbulence:** Internal friction (within the air) is caused by uneven surfaces.
- **Local factors:** Sometimes buildings and other local features can influence wind patterns, and either slow down the wind or create a 'wind tunnel' effect.

Knowledge check 17

How does the Coriolis effect work differently in the northern and southern hemispheres?

Exam tip

Make sure that you understand and can fully explain the way that atmospheric pressure works to create wind that transfers heat.

Summary

- The global energy balance is an open system with insolation, storage of heat and some heat lost back into space.
- Surplus heat is transferred around the globe horizontally (through ocean currents, winds and hurricanes) and vertically (through radiation, convection and conduction).
- Ocean currents play an important role in moving heat energy around the world.
- Wind occurs due to differences in atmospheric pressure in an area (the pressure gradient). The bigger the difference in pressure, the stronger the wind will be.
- Wind direction is affected by three controlling factors: pressure gradient, the Coriolis effect and friction.
- Surface pressure belts cause the movement of air from areas of high pressure to areas of low pressure.
- The general circulation of the atmosphere is dominated by the tri-cellular model. This helps to explain how the Earth can redistribute energy received at the Equator towards the poles.
- The jet stream is a moveable ribbon of fast flowing air at high altitudes in the atmosphere, which can have a big impact on weather.
- The global temperature pattern is affected mostly by latitude, but also by ocean currents, continentality, altitude, seasonality and prevailing winds.

Weather in the British Isles

The British Isles is a good example of a geographic area within the mid-latitudes, where there is a mixing of air between the Ferrel cell and the Polar cell (at the polar front/jet stream). This means that the weather changes constantly.

The formation of precipitation

Orographic (relief) rainfall

As a parcel of air is moved (by wind) from the sea towards the land it is often forced upwards as the hills or mountains cause a physical barrier. As the air is forced to rise, it expands, cools adiabatically and the **relative humidity** increases. At 100% relative humidity the **dew point** is reached, resulting in **precipitation**.

Convectional rainfall

Convection currents take place when the air above the surface of the Earth is heated by the direct heating of the sun. The air is forced to expand, it cools adiabatically and relative humidity increases. At 100% relative humidity the dew point is reached and the air condenses, producing towering cumulonimbus clouds. These create unstable air, which causes thunderstorms and in some cases creates the conditions for a hurricane to form. Rain events can be short but intense.

Cyclonic (frontal) rainfall

Cyclonic rain occurs when two bodies of air (air masses) meet. This can happen at the polar front when warm air from the south meets with cold air from the north. At a front the air can become very unstable, because the warm and cold air cannot mix. The warm air starts to rise up and over the colder (dense) air. As the air expands, it cools adiabatically, leading to precipitation. Frontal rain is often associated with the development of depressions.

The formation of mid-latitude weather systems

What are air masses?

An air mass is a large parcel of air that can be thousands of kilometres wide. It remains in one location for a long period of time and picks up the area's temperature and moisture characteristics. The air mass can then move steadily across an area, bringing constant temperature and humidity conditions.

The five air masses that affect the British Isles (Figure 36) are:
- tropical maritime (Tm) — warm and moist
- tropical continental (Tc) — warm and dry
- polar maritime (Pm) — cold and (fairly) moist
- polar continental (Pc) — cold and dry
- arctic (A) — cold, but with snow.

Relative humidity The amount of water vapour in the air expressed as a percentage of the total amount of water vapour air can hold at that temperature.

Dew point The temperature to which a parcel of unsaturated air must be cooled in order to become saturated (full up with water vapour).

Precipitation Any form of water moving through the air towards the surface of the Earth. It includes drizzle, rain, fog, mist, dew, hoar frost, hail, sleet and snow.

Knowledge check 18

For each of the three main causes of precipitation, use an annotated diagram to help explain how they allow precipitation to develop.

Exam tip

Make sure that you have a detailed understanding of how each of the causes of precipitation works. They are all based on air being forced to rise.

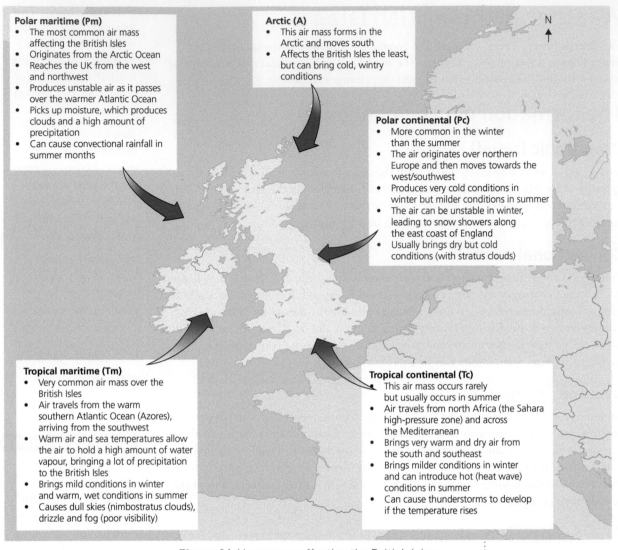

Polar maritime (Pm)
- The most common air mass affecting the British Isles
- Originates from the Arctic Ocean
- Reaches the UK from the west and northwest
- Produces unstable air as it passes over the warmer Atlantic Ocean
- Picks up moisture, which produces clouds and a high amount of precipitation
- Can cause convectional rainfall in summer months

Arctic (A)
- This air mass forms in the Arctic and moves south
- Affects the British Isles the least, but can bring cold, wintry conditions

Polar continental (Pc)
- More common in the winter than the summer
- The air originates over northern Europe and then moves towards the west/southwest
- Produces very cold conditions in winter but milder conditions in summer
- The air can be unstable in winter, leading to snow showers along the east coast of England
- Usually brings dry but cold conditions (with stratus clouds)

Tropical maritime (Tm)
- Very common air mass over the British Isles
- Air travels from the warm southern Atlantic Ocean (Azores), arriving from the southwest
- Warm air and sea temperatures allow the air to hold a high amount of water vapour, bringing a lot of precipitation to the British Isles
- Brings mild conditions in winter and warm, wet conditions in summer
- Causes dull skies (nimbostratus clouds), drizzle and fog (poor visibility)

Tropical continental (Tc)
- This air mass occurs rarely but usually occurs in summer
- Air travels from north Africa (the Sahara high-pressure zone) and across the Mediterranean
- Brings very warm and dry air from the south and southeast
- Brings milder conditions in winter and can introduce hot (heat wave) conditions in summer
- Can cause thunderstorms to develop if the temperature rises

Figure 36 Air masses affecting the British Isles

The formation of a depression

Depressions are areas of low atmospheric pressure that can produce cloudy, rainy and windy weather conditions. The depressions that affect the British Isles are formed out in the Atlantic Ocean when cold polar maritime air from the north moves south and meets warm, tropical maritime air that is moving north. The place where the cold, ex-polar air and the warm, relatively moist air from the tropics meets is the **polar front**. The lighter, warm air will start to rise up and over the colder, denser air from the north. The rising warm air will mean that atmospheric pressure is reduced (leaving a low pressure). This leads to a disturbance called a **baroclinic instability**, which continues to develop into a front (Figure 37).

Knowledge check 19

Describe the key characteristics of the dominant air mass that affects the British Isles.

As the air rises at the polar front the winds in the upper atmosphere are organised into a fast flowing jet stream called the **polar front jet stream**. The winds associated with the polar front jet stream are very important as they influence the track of mid-latitude weather systems. If the jet stream lies to the north of the British Isles, the depression will move north of Scotland and the rest of the British Isles will not experience the bad weather. However, if the jet stream moves south and lies over the British Isles, the low-pressure systems will pass over the British Isles and bring cloud, wind, precipitation and changeable weather.

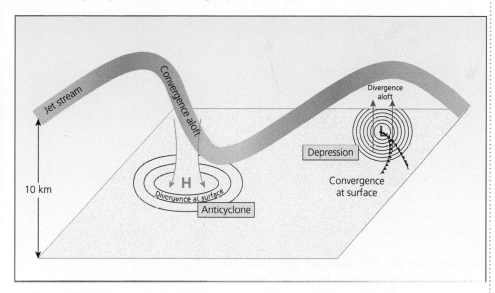

Figure 37 How the polar front jet stream influences depressions and anticyclones

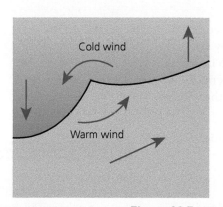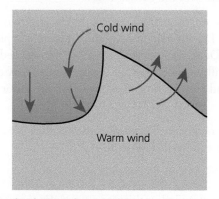

Figure 38 Formation of a depression

Exam tip

You need to know the subtle differences between each of the different air masses.

Weather associated with a depression

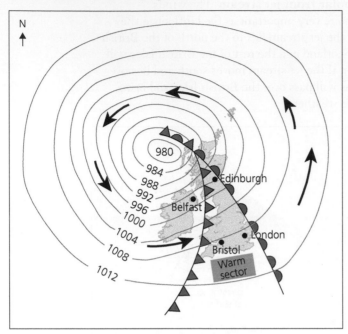

Figure 39 Depression over the British Isles

Depressions create a very distinctive pattern as they pass over the British Isles (Figure 39). Most move from the SW towards the NE. The full lifespan of a depression from formation to decay is around 5 days, but it normally takes around 24 hours for a depression to pass over the British Isles.

As the cold air gives way to the warm air from the southwest (the warm sector), changes to the weather result. During a depression the pressure is typically low (below 1,000 mb) and falling. Depressions often encourage wind speeds to increase — isobars on weather maps will be close together. Temperatures vary depending on the type of air passing overhead and the time of year — they are lower in the winter months.

The depression creates a wide range of clouds and corresponding precipitation that varies as the depression passes overhead (Figure 40).

Exam tip

You might be asked to show detailed understanding of the formation and structure of a depression, so make sure that you understand this fully. Questions often ask you to draw a diagram of the passage of a depression.

Knowledge check 21

Describe and explain the sequence of events that are associated with a typical depression passing overhead.

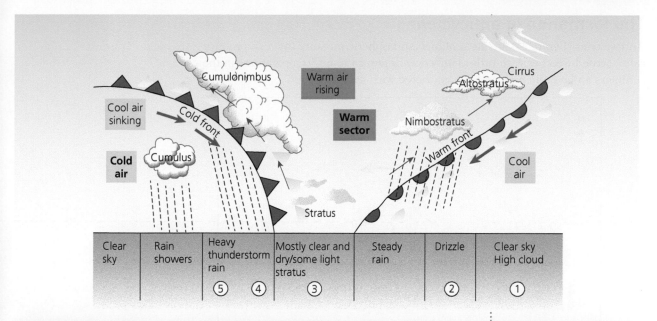

	⑤Cold sector	④Passage of cold front	③Warm sector	②Passage of warm front	①Ahead of warm front
Pressure	Rising	Starts to rise	Steadies	Low	Starts to fall steadily
Temperature	Cold temperatures	Temperatures start to drop quickly	Warm	Temperatures rise	Quite cold
Cloud cover	As cumulonimbus clouds pass they give way to 'fair weather' cumulus clouds	Clouds thicken quickly with some towering cumulonimbus	Some clear skies with some light stratus clouds	Clouds are lower and thicken (altostratus and nimbostratus)	Cloud base drops and clouds get thicker (cirrus and altostratus)
Wind speed and direction	Wind speeds again start to decrease as depression passes (NW)	Wind speeds increase, sometimes to gale force (SW to NW)	Calmer conditions (SW)	Strong winds (SE to SW)	Wind speed increases and wind direction changes (S)
Precipitation	Some rain showers, which gradually ease	Heavy thunderstorms with a chance of thunder and lightning; heavy rain and even hail	Rain, turning to drizzle and then dry conditions for a short period	Some sustained drizzle due to nimbostratus clouds	Little at first

Figure 40 Passage of a depression

The formation of anticyclones

Anticyclones (Figure 41) are areas of high atmospheric pressure that can produce calm, settled weather with little cloud cover or precipitation. Temperatures in the summer can be quite high and are associated with 'good' weather.

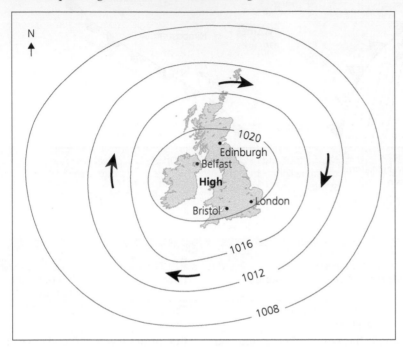

Figure 41 Typical synoptic chart for an anticyclone

The high-pressure system brought in by an anticyclone usually comes from the south of the UK. **Tropical continental air masses** arrive from areas of high pressure and bring warm and dry air into the British Isles. As the air sinks from a high altitude, condensation cannot take place. Anticyclonic conditions can also occur when the polar continental air mass develops high pressure and sinking air conditions.

Anticyclones can also be caused due to a convergence aloft in the polar front jet stream. As the convergence aloft builds pressure, the air starts to sink, which will create an area of high pressure at the Earth's surface.

Anticyclones can move into an area for a few days or sometimes for a longer period of time, blocking out any other weather system.

Weather associated with anticyclones

The weather produced by an anticyclone working in the summer can be very different from what happens in the winter (Table 6).

Table 6 Weather associated with anticyclones

	Summer anticyclone	Winter anticyclone
Pressure	High and increasing (over 1,000 mb)	High and increasing (over 1,000 mb)
Temperature	Temperatures can be warm — air from the warm south allows temperatures on hot sunny days to go above 24°C	Winter temperatures are much lower — the sun is low in the sky and there is less heat
Cloud cover	Sinking air allows for very settled/stable conditions and during the day temperatures increase; however, clear skies can lead to cold nights	As in summer, but lack of cloud cover at night can cause temperatures to go below freezing and lead to frost and ice
Wind speed and direction	Isobars are far apart —conditions are calm and any wind moves in a clockwise direction	As in summer
Precipitation	Very little, if any, precipitation, but sometimes dew or fog in the evening; thunderstorms can develop if heat leads to convectional rainfall conditions	Very little, if any, precipitation but sometimes the lack of cloud cover can cause freezing fog and frost/icy conditions

Knowledge check 22

Describe the main differences and similarities between winter and summer anticyclones.

Exam tip

Make sure that you know and understand the main differences as to how an anticyclone might operate in the summer and the winter. The differences are subtle but they can trigger very different weather features.

Interpreting weather systems that affect the British Isles

Weather forecasters have used **synoptic charts**/surface pressure charts for many years to enable them to predict the way that the weather over an area will change in the coming days. More recently, satellite imagery allows meteorologists to observe cloud patterns.

Using surface pressure charts

Surface pressure charts (synoptic charts) show the details of atmospheric pressure and weather fronts. Atmospheric pressure is displayed using **isobars** (a line that joins a place with equal pressure). Isobars are usually drawn every 4 mb (with the air pressure always written along the isobar). When isobars are close together the pressure gradient is increased, which means that winds will be stronger.

A weather analysis chart is drawn to show the observed state of the weather. This is updated every 12 hours.

Fronts are also shown on the chart. A weather front is a boundary between two air masses. There are three different types of front.

- **A cold front:** a cold front on a surface pressure chart shows that a block of cold air is moving. This means that the weather will change as the cold air moves in — a narrow belt of rain and clouds will indicate the arrival of the front. When the front passes, the weather will settle again and become bright and clear. Atmospheric pressure usually falls before the front passes, and then rises afterwards.

■ **A warm front:** warm fronts can bring a bank of cloud with precipitation as the block of warm air moves overhead. As the front passes, the rain can get quite heavy and will then reduce. The warm front does not always bring warm air. Pressure will fall quickly before the front continues to move.

■ **An occluded front:** this occurs when warm air has already started to rise up over cold air and brings the same weather as at a cold front with less cloud and rain. However, sometimes it can bring a band of thunderstorms.

Knowledge check 23

Describe the main differences between the different fronts on a surface pressure chart.

Exam tip

Make sure that you practice reading surface pressure charts and satellite images. You could go to www.metoffice.gov.uk to look up the most recent surface pressure charts and compare these to the satellite images and the forecasts for the week ahead.

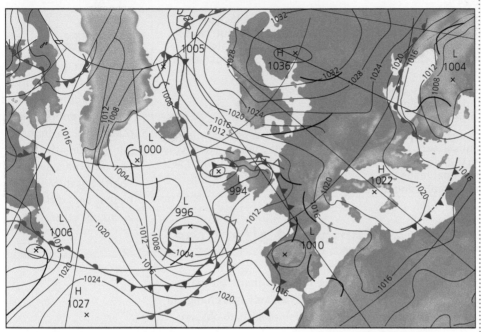

Figure 42 Surface pressure chart (12 September 2015) (contains public sector information licensed under the Open Government Licence v3.0)

Table 7 Symbols used on a surface pressure chart

	Cold front
	Warm front
	Occluded front (or 'occlusion')
———**996**———	**Isobars** (the max./min. pressure in an area is marked as H (high) or L (low))
——————	**Trough** (a long area of low pressure which usually indicates increases in cloud and precipitation)

One simple way of starting to interpret a surface pressure chart is to work out the wind direction using **Buys Ballot's Law**. This states that if, in the northern hemisphere, a person stands with their back to the wind, pressure is lower on the left hand than on the right. This helps to remind us that wind blows anticlockwise around a depression and clockwise around an anticyclone.

Figure 43 opposite shows a version of the surface pressure chart shown in Figure 42 above, but it has been annotated with some of the basic weather features.

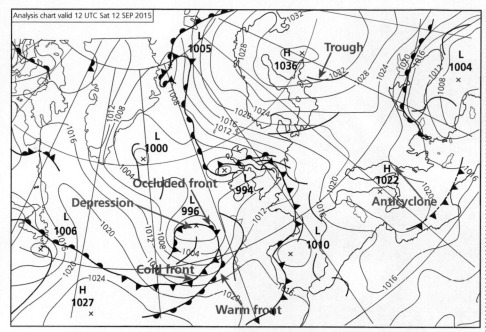

Figure 43 Surface pressure chart (12 September 2015), with annotation (contains public sector information licenced under the Open Government Licence v3.0)

Using satellite imagery

Figure 44 Satellite image of a depression over northwest Ireland on 5 June 2015

Figure 45 Satellite image of an anticyclone over Europe on
21 April 2015

Two types of satellite are used to observe the weather. **Visible images** (Figures 44 and 45) record any visible light from the sun that is reflected back from the top of clouds and land/sea surfaces. These are better for seeing low cloud but can only be used in daylight hours. **Infrared images** will show any infrared radiation that is given off from clouds or the sea. Lighter areas of cloud show where fronts and shower clouds can be found.

On satellite images, anticyclones are recognisable due to their lack of cloud. There will be more cloud obvious when a depression is present. Depressions are often presented on satellite pictures with a swirl of cloud. Frontal systems might have a wishbone shaped area of cloud that moves out from the centre of a depression.

Summary

- There are three main types of precipitation: orographic, convectional and cyclonic (frontal).
- Air masses control the formation of mid-latitude weather systems.
- An air mass is a large parcel of air that stays in one location and picks up the area's temperature and moisture characteristics. The five air masses that affect the British Isles are:
  - tropical maritime (Tm) — warm and moist
  - tropical continental (Tc) — warm and dry
  - polar maritime (Pm) — cold and (fairly) moist
  - polar continental (Pc) — cold and dry
  - arctic (A) — cold but with snow
- Depressions are formed when polar maritime air from the north moves south and meets warm, tropical maritime air that is moving north. This causes a baroclinic instability, which creates a front.
- As a low-pressure system passes overhead, the weather changes. The different weather elements are affected in different ways as the cold air, warm sector and cold air pass.
- Anticyclones are areas of high atmospheric pressure that produce calm, settled weather with little cloud or precipitation.
- Anticyclones can move into an area for a few days or they might remain for a longer period of time and block any other weather system.
- Anticyclones can cause slightly different weather features in summer and winter.
- The weather systems that affect the British Isles can be interpreted using surface pressure charts and satellite imagery. This helps us to forecast what weather might be coming over the next few days.

Global weather issues

The El Niño Southern Oscillation and La Niña

El Niño (Southern Oscillation, or ENSO) (Figure 46(b)) is a weather phenomenon that involves an unusual warming up of the surface of the Pacific Ocean, mostly around the Tropics. It has a huge impact on world weather patterns and sometimes the process is reversed (La Niña). The National Oceanic and Atmospheric Administration (NOAA) definition for El Niño is for a 3-month warming of at least 0.5°C above average in the eastern tropical Pacific Ocean.

In a **normal year** (Figure 46(a)), the heat transfers in the atmosphere take place to move heat from the Equator and subtropics using the trade winds from South America towards Australia. This takes the warm water west and allows a cold current of water to run along the Pacific coast of South America (the Peru current).

Every 3–8 years this circulation pattern breaks down. The trade winds weaken and warm water moves from the east to the west, stopping the movement of the colder water. This allows a warm ocean current to exist off the South American coast. It usually arrives around Christmas time and so was named El Niño (the Christ child).

La Niña (Figure 46(c)) is the opposite of El Niño. In this the eastern trade winds will blow stronger than usual, which pushes the warm ocean water west, creating a very low pressure over Australia. This causes descending air over South America and can produce very high pressure, which can lead to drought conditions.

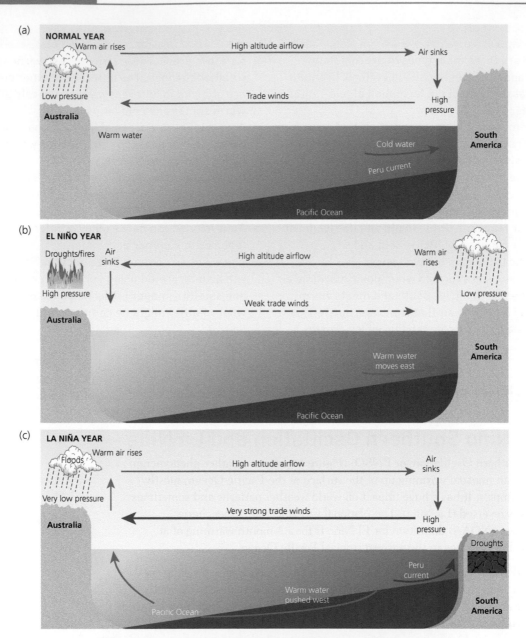

Figure 46 The changes brought by El Niño and La Niña events

Table 8 The main features of El Niño and La Niña events

	Normal year	El Niño year	La Niña year
Features	Warm air is moved towards the west due to trade winds. This creates a low pressure in Australia and a high pressure in South America.	The trade winds weaken, which leads to a reversal from west to east. Warm ocean water moves eastwards and subdues the Peru current, bringing warm water to the South American coast.	The trade winds blow more strongly than usual and can push warm oceanic water towards Australia, creating a very low pressure. Descending air over South America leaves high pressure.
Impact on global wind	Normal pattern of global circulation. Trade winds move heat from east to west with a reverse flow at high altitudes.	As the trade winds have been weakened and reversed this causes air to rise over South America. The high-altitude airflow is reversed and air descends over Australia, leaving a high pressure weather system.	The strong trade winds blow across the ocean. The high-altitude air flows from west to east and leaves high pressure over South America.
Impact on rainfall patterns	Normal pattern of circulation causes rising air and depressions to form in Australia, with descending air over South America bringing settled weather and reduced rainfall. Normal weather patterns.	In South America the normal high pressure is replaced with low pressure. This can increase temperatures by between 6°C and 10°C. The rising air brings heavy rainfall along the usually dry coast of Peru. South America can sometimes experience flash floods, though in Australia the high-pressure system can lead to long periods of drought that can cause bushfires. There is a noticeable increase in floods and snowfall in North America and a reduction of hurricane events in the Caribbean during El Niño events.	Rising air over Australia causes depressions to form and the influx of warm air can cause flooding. The sinking air over South America can bring extreme drought to already arid areas. Rainfall will be much higher over the western Pacific (leading to flooding) and much lower over South America (leading to drought conditions). In addition, South Africa can experience floods while more northern parts of Africa can experience drought. Some evidence suggests that more hurricanes will form in the Caribbean.

Since its discovery, scientists have identified that El Niño has killed fish along the South American coast (1972), and caused extreme drought in California (1977), extreme global weather such as the drought in Africa (1983) and extreme floods across South America (1983 and 1999). The arrival of El Niño has also been linked to the increased spread of water-based diseases such as cholera, dengue fever and malaria.

The formation and structure of hurricanes, tropical cyclones or typhoons

Hurricanes are sometimes known as tropical revolving storms, tropical cyclones or typhoons. They are **severe tropical depressions** where the wind speeds go above 64 knots (114 km h^{-1}). They happen when the pressure is extremely low (below 970 mb), sometimes as low as 880 mb.

The formation of hurricanes

Hurricanes happen most often in the late summer and autumn. They always track from **east to west** and always spin away from the Equator. Scientists still do not entirely understand the actual mechanics of how tropical cyclones start to form, but the following factors are thought to be necessary:

Table 9 The formation of hurricanes

Factor	Explanation
Ocean temperatures: temperatures need to be 27°C to a depth of 60 m.	If temperatures are this high, it gives a sufficient source of heat to cause the convectional uprising of air needed for the hurricane to form.
Seasonality: hurricanes form in late summer to late autumn (that is from July to November in the northern hemisphere).	By this stage of the year the sea has had enough time to heat up sufficiently, to 27°C to a depth of 60 m.
Pre-existing disturbed weather pattern: for example, a wave in the easterly trade winds off the coast of east Africa as the hot air from the Sahara meets the cooler air from the Gulf of Guinea.	This provides the initial energy and spin required to start the process of hurricane formation.
Latitude: hurricanes will only form between 5° and 20° N/S.	The Coriolis effect is too weak between the Equator and 5° N/S to produce the spiralling of winds needed to form a hurricane.
High humidity levels: (about 80% relative humidity), which are found over oceans as the water evaporates.	High levels of humidity are needed to cause and sustain hurricanes because this brings the air close to saturation. This means that, as the air rises and cools adiabatically, it quickly reaches dew point. When condensation then occurs this releases latent heat, which heats the surrounding air, helping to sustain convectional rising.

The structure of hurricanes

Hurricanes can be massive weather systems spanning over 1,000 km and reaching up to 12 km into the troposphere/atmosphere. The eye is around 50 km wide. The structure of a hurricane is shown in Figure 47. Hurricanes only start to lose power, intensity and wind speed when they lose their energy supply (i.e. warm seawater). This means that hurricanes decrease in strength when they pass over colder water or hit landfall (but they can still be powerful storms).

The impacts of hurricanes

The main impacts of a hurricane are caused by extreme wind speeds and flooding from swells which can actually bring huge amounts of damage and destruction to low, coastal environments. Every hurricane/tropical cyclone/typhoon will bring its own problems and local factors often determine the amount and type of damage.

Exam tip

You need to know the difference between the conditions needed for a hurricane to form and the structure and features of a hurricane when it is fully formed.

Knowledge check 24

Describe and explain some of the key factors required in the formation of a hurricane.

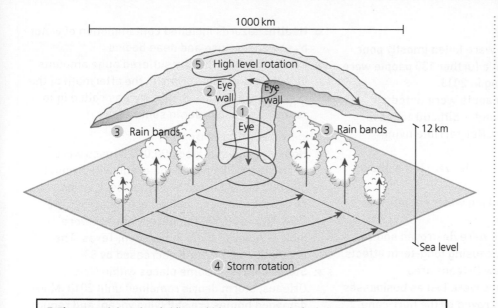

1 The eye: air here is subsiding, which gives a period of relative calm

2 The eye wall: air here is rising quickly due to strong convection currents, and surface air is drawn upwards quickly. Cumulonimbus clouds form and stretch up to the top of the troposphere (12 km)

3 Rain bands: a series of rain bands forms on either side of the eye walls as there are further movements of air upwards due to convection currents

4 Storm rotation: in the northern hemisphere the storm rotates, with surface winds spiralling in towards the eye (in an anticlockwise direction)

5 High-level rotation: at higher levels the air moves in the opposite direction (clockwise in the northern hemisphere)

Figure 47 Structure of a hurricane

Case study

Hurricane Katrina (USA), 29 August 2005

Hurricane Katrina was a category 5 hurricane. It started to form in late August in the Atlantic Ocean. The hurricane tracked to the west and moved across the Bahamas and through Florida on 25 August. It continued to generate energy in the warm seas of the Gulf of Mexico and moved north, hitting New Orleans on 29 August. On 28 August the mayor of New Orleans, Ray Nagin, when announcing the first ever evacuation of the city, noted that Katrina was 'a storm that most of us have long feared'.

What were the main hazards?

■ **High wind speeds:** as a category 5 hurricane, Katrina sustained wind speeds of over

$250\,\mathrm{km\,h^{-1}}$. At its height, winds of $282\,\mathrm{km\,h^{-1}}$ were recorded.

■ **Rainfall:** the hurricane brought a large amount of rainfall, which battered the coast for a sustained period before and after the storm had passed. Peak rainfall levels were measured at over 250 mm over a 24-hour period.

■ **Storm surge:** an 8.5 m-high surge of water caused by high winds and large waves pushed water inland, flooding the New Orleans flood levée system. Up to 80% of the city became flooded and flood waters were pushed 19 km inland.

The impact on people

- **Deaths:** 1,836 people were killed (mostly poor or elderly people), but a further 135 people were still classed as missing in 2013.
- **Homeless:** 500,000 people were listed as homeless. People found it difficult to leave the area and had to suffer terrible living conditions.
- **Electricity:** 3 million people were left without electricity.
- **Water supply:** all drinking water supplies were polluted with sewage.
- **Education:** 18 schools were destroyed and 74 were badly damaged, causing long-term effects to schooling in the New Orleans area.
- **Economic:** 230,000 jobs were lost as businesses were damaged or destroyed by the hurricane event. Many ports and oil platforms in the Gulf of Mexico were damaged and had to close. Damage was estimated at $108 billion — the costliest hurricane in US history.

- **Health:** hazards included contamination of water by sewage, refuse and dead bodies.
- **Law and order:** the city suffered huge amounts of looting and other crime in the aftermath of the hurricane, and the US military was called in to restore law and order in some areas.

The impact on property

- **Housing:** an estimated 300,000 houses were destroyed by the hurricane event, with around 500,000 people listed as homeless. Katrina redistributed over 1 million people across the USA. Over 100,000 temporary houses were built — many of them in Houston, Texas. The population of the state decreased by 8%.
- **Storm debris:** in some places within New Orleans, storm debris remained until 2010. Many damaged homes are still lying vacant and require serious rebuilding work over a decade later.
- **Rebuilding:** 6 months after the hurricane, the city centre in New Orleans was still without a working sewage system, or gas and electricity supply.

Evaluation of management strategies to reduce impact on people and property

Case study

Hurricane Katrina (USA), 29 August 2005

Positive measures

- **Early warning systems:** the USA has a very organised system for monitoring and predicting the track of hurricanes and major storms, called the National Hurricane Centre (NHC). It was able to give accurate forecasts about the track of the hurricane and warn where the major impacts might be felt.
- **Evacuation plan:** New Orleans had a comprehensive disaster and evacuation plan in readiness for an event like Katrina. The issue was that many people chose to ignore the warnings.
- **Emergency service preparation:** the government authorities had practised their responses to a storm event like this. They had

created a fictional hurricane called 'Hurricane Pam' in 2004 and modelled the different responses needed in order to cope with a category 3 hurricane. However, one failure of this exercise was not to consider the impact of levée collapse, and yet it estimated that 60,000 people would be killed.

Negative measures

- **FEMA and disaster planning:** the Federal Emergency Management Agency (or FEMA) began to assist the local authorities before the storm and continued for more than six months after the storm. A year on, the former FEMA chief noted that 'there was no plan' for dealing with the aftermath of Katrina.

- **Evacuation planning:** 25% of the New Orleans population did not own a car, so leaving the city was difficult. There were huge tailbacks as people tried to flee the city. Mayor Nagin was criticised for delaying his emergency evacuation order until 19 hours before the hurricane made landfall.
- **Stranded people:** shortly after the hurricane had taken place media images began to appear of people who were left stranded by flood waters, without food, water or shelter, as death rates started to mount up. Many felt that the government response was slow and delayed in getting the right aid to the right people.
- **Emergency service preparation:** the breaching of the levées meant that mass flooding became a huge issue in New Orleans. Aid could only enter the areas of need using overland routes. The Louisiana Superdome was the 'shelter of last resort' for some 26,000 people who could not escape the city.
- **Levée protection:** the US Corps of Engineers had constructed the levées that surrounded the city of New Orleans and protected it from the Mississippi River and the Gulf of Mexico. Much of the city was up to 10 m below sea level. The levées were built to survive a category 3 hurricane event but not a category 5. The US Congress had turned down plans to upgrade the defences to withstand a category 5 event and the budget for the US Corps of Engineers had been under pressure in the 10 years leading up to Katrina.
- **Building design and insurance:** any building in a potential hurricane zone must be built to tight regulations. Many of the buildings in New Orleans were built to withstand the hurricane-force winds, but this did nothing to limit the impact of flooding in the area.

Summary

- El Niño is a weather phenomenon that involves a warming of the surface of the Pacific Ocean around the tropics. Heat is moved from the Equator (from South America to Australia) and a cold current then runs along the Pacific coast of South America.
- La Niña is the opposite of El Niño. The eastern trade winds will blow more strongly than usual, which pushes the warm ocean water west, creating very low pressure over Australia and a high pressure over South America.
- Hurricanes are storms with high winds (above 114 km h^{-1}) and extremely low pressure (below 970 mb).
- Hurricanes always track from east to west and spin away from the Equator. They usually happen in the late summer and autumn.
- There are five key factors necessary for hurricane formation.
- The structure of a hurricane is very different from any other weather system. Features include the eye, eye walls, rain bands and storm rotation.
- Hurricanes are huge systems spanning over 1,000 km and reaching up to 12 km into the troposphere.
- Hurricane Katrina was a category 5 hurricane that hit the coast of the USA on 29 August 2005, causing a huge amount of damage to the city of New Orleans.
- Katrina had a big impact on the people and property in the southern USA — killing 1,836 people and destroying 300,000 houses, with damage costing an estimated $108 billion.
- Many of the protective measures and management strategies put in place to reduce the loss of life and damage to property in the New Orleans area can be evaluated both positively and negatively.

Questions & Answers

The AS Unit 1 Geography paper contains six questions:

Unit structure

	Compulsory?	Marks (out of 75)	Exam timing (out of 75 minutes)
Section A			
Q1 Rivers: short structured question	Yes	15	15
Q2 Ecosystems: short structured question	Yes	15	15
Q3 Atmosphere: short structured question	Yes	15	15
Section B			
Q4 Rivers: extended question	Answer two from questions 4, 5 or 6	30 marks = 15 marks for each question	30 minutes = 15 minutes for each question
Q5 Ecosystems: extended question			
Q6 Atmosphere: extended question			

Examination skills

As with all AS exams there is little room for error if you want to get the best grade. Gaining a grade A is not easy in AS geography so you need to ensure that every mark counts.

The examination papers for AS Unit 1 and Unit 2 are both 1 hour and 15 minutes. There are 75 marks available on each, which means that you get 1 mark per minute to work your way through the paper. The main reason why so many students struggle with this paper is that they fail to manage their time appropriately and as a consequence they do not have enough time left to answer the essays at the end in sufficient detail. If you find that you have time left over in this exam, the chances are that you have done something wrong.

Exam technique

Students often find it difficult to break an exam question down into its component parts. On CCEA exam papers, the questions are often long and difficult to understand, so you need to work out what the question is asking before you move forward.

Command words

To break down the question properly, get into the habit of reading the question at least **three** times. When you do this it is sometimes a good idea to put a circle round any command or key words that are being used in the question.

A common mistake is failing to understand the task that the question is setting. There is a huge difference between an answer asking for a discussion and one asking for an evaluation.

The main command words used in the exam are as follows.

- **Compare:** what are the main differences and similarities?
- **Contrast:** what are the main differences?
- **Define:** state the meaning (definition) of the term.
- **Describe:** use details to show the shape/pattern of a resource. What does it look like? What are the highs, lows and averages?
- **Discuss:** describe and explain. Argue a particular point and perhaps put across both sides of this argument (agree and disagree).
- **Explain:** give reasons why a pattern/feature exists, using geographical knowledge.
- **Evaluate:** look at the positive and negative points of a particular strategy or theory and finish with a concluding statement about what you think is the stronger argument.
- **Identify:** choose or select.

Structure your answer carefully

Sometimes the longer questions on exam papers can prevent students from achieving maximum marks. Questions which are marked from 6 to 15 marks will usually be marked using three Levels. Later in this section we will look at some questions and give more guidance about how you should structure your answers.

One simple approach to consider is drawing up a brief plan for your answer so that you know where it is going and how you will cover all of the main aspects of the question. For example, you could draw a box to illustrate each element needed within an answer and fill each one with facts and figures to support the answer, using the marking guidance to help you work out how much time to spend on each section.

Show your depth of knowledge of a particular place/case study

The extended writing questions on the exam paper are usually focused on giving the student the opportunity to apply knowledge and understanding of case study material to a particular question. It is really important to show what you know here.

Examiners are looking for specific and appropriate details, facts and figures to support your case. The better you know and understand your case studies, the higher the marks you can potentially achieve.

About this section

A practice test paper with exemplar answers is provided. This will help you to understand how to construct your answers in order to achieve the highest possible marks.

Examiner comments

Some questions are followed by brief guidance on how to approach the question (shown by the icon ⓔ). Student responses are followed by examiner's comments. These are preceded by the icon ⓔ and indicate where credit is due. In the weaker answers, they also point out areas for improvement, specific problems and common errors such as lack of clarity, weak or non-existent development, irrelevance, misinterpretation of the question and mistaken meanings of terms.

■Section A

Question 1 Rivers: short structured questions

(a) (i) With the aid of a well-annotated diagram, describe and explain the role of deposition in the formation of a floodplain.

(6 marks)

ⓔ **Level 3 (5–6 marks):** The answer uses a well-annotated diagram and specialist terminology to explain the formation of the floodplain and associated river processes.

Level 2 (3–4 marks): The answer might use a less-well-annotated diagram of the feature and/or a less detailed explanation of its formation. Alternatively, an accurate and detailed explanation with no diagram might be marked at this level.

Level 1 (1–2 marks): A very limited response that fails to offer any meaningful explanation of the processes at work.

Student answer

(a) (i) There are two main sources of deposition. The first are called point bar deposits, which are deposited on the lower energy inside bends of migrating meanders. As the meanders weave across the floodplain, they leave these point bar deposits all across the floodplain. Second, if a river floods onto the floodplain, it experiences a lot of friction as it comes into contact with the ground and the vegetation on it. This friction slows the velocity of the river, causing deposition to occur. The largest particles need most energy to transport them, so they are deposited first (forming levées along the edge of the channel), while the smallest particles are carried furthest across the floodplain.

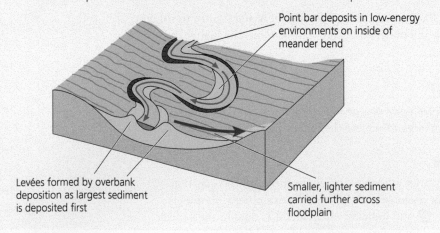

Point bar deposits in low-energy environments on inside of meander bend

Levées formed by overbank deposition as largest sediment is deposited first

Smaller, lighter sediment carried further across floodplain

ⓔ **6/6 marks awarded.** This detailed answer shows good understanding of the depositional processes at work in the formation of a floodplain. There is good use of terminology and a well constructed and annotated diagram is included.

(ii) Briefly explain how deposition contributes to the formation of another river feature apart from floodplains. (3 marks)

e Award 1 mark for an answer that identifies another river feature and 2 marks if it outlines how deposition occurs to help form the feature, with appropriate use of terminology. Award 1 for an answer that lacks detail in explanation of the contribution that deposition plays.

> (ii) Deposition is also significant in the formation of meanders. On the inside of the river bend, velocities are lower here than on the outside. This means that, as river energy drops here, sediment is deposited to form a gently sloping bank made up of point bar deposits.

e **3/3 marks awarded.** A relevant river feature is identified and the role deposition plays is discussed, with good use of appropriate terminology.

(b) Study Resource 1, which shows the Hjulström curve. After a rainstorm, the discharge in a river falls from 100 cm s^{-1} to 1 cm s^{-1}. Use the Hjulström curve to describe and explain how this impacts on the river's load. (6 marks)

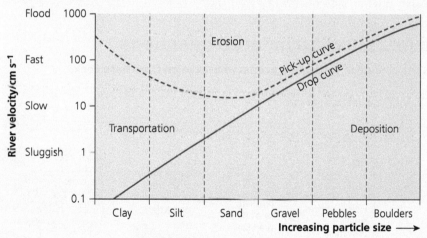

Resource 1 Hjulström curve

e **Level 3 (5–6 marks):** A detailed and thorough answer, which makes effective use of the resource to discuss the changes over time.

Level 2 (3–4 marks): A general but accurate answer, which discusses how the load changes over time. Use of the resource may be restricted.

Level 1 (1–2 marks): A limited answer, which fails to address the changes over time and/or makes no meaningful reference to the resource.

(b) At the start of the time period, when the velocity is at 100 cm s^{-1}, the river has enough energy to transport a large range of sizes of load, from clay up to gravel-sized particles. However, during this time period, as velocity drops from 100 cm s^{-1} to 1 cm s^{-1}, the river's energy drops and so its carrying capacity also drops. This means that the river begins to deposit more and more load, starting with the largest particles — gravel. The main reason for this is that gravel is larger and so requires more energy to transport than smaller particles such as sand. As velocity drops further, however, the smaller particles begin to be deposited: sand is deposited at around 10 cm s^{-1} and silt at around 1 cm s^{-1}. Below 1 cm s^{-1}, the smallest clay particles continue to be transported, as they are so light.

ⓔ **6/6 marks awarded.** This detailed and thorough answer clearly addresses the changes over time. Resource use is detailed, with plenty of figures quoted, and the changes are clearly explained.

ⓔ The focus of this question is on how the transportational *and* depositional processes change during the time period. As discharge and therefore river energy fall, the load begins to be deposited, starting with the largest particles.

Question 2 Ecosystems: short structured questions

Study Resource 2, which shows the change in vegetation primary succession from bare rock.

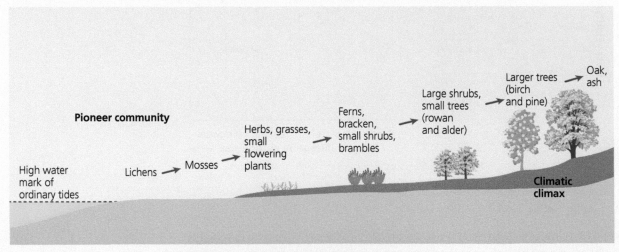

Resource 2 Primary succession on bare rock

(a) Using examples, distinguish between primary and secondary succession. (4 marks)

ⓔ 2 marks are for demonstrating clear understanding of the contrasting definitions and 2 (×1) marks are for an example of each.

Student answer

(a) Primary succession happens when plants colonise and grow on a surface that did not have vegetation on it before, such as bare rock on a newly formed volcanic island. In contrast, secondary succession occurs on surfaces where the vegetation has been destroyed, such as when fire creates a clearing in a forest.

e **4/4 marks awarded**. This communicates the clear distinction between the two processes and quotes good examples. This is a case of the student reading the question carefully. See the few extra seconds it takes to read the question carefully as a vital investment and not a waste of your time.

(b) Explain **one** soil change and **one** microclimate change that occur to produce the succession shown in Resource 2.

(4 marks)

e As succession occurs, the abiotic environment is progressively changed. You should explain one change to the *soil* (e.g. depth, humus, moisture, pH) and one to the *microclimate* (e.g. wind speed, temperature).

(b) The soil becomes deeper due to the addition of more organic matter from the biomass and increased weathering of the bedrock to provide mineral content. Also, the microclimate improves, with the vegetation causing frictional drag on the wind and reducing surface wind speeds.

e **4/4 marks awarded**. This is clearly focused on the requirements of the question and shows a clear understanding of the two changes selected.

(c) With reference to a named small-scale ecosystem you have studied, discuss the role of decomposers in the ecosystem.

(4 marks)

e 3–4 marks are awarded for an answer that identifies decomposers and explains their role. An ecosystem must be named to achieve these marks. An answer that does not fully explain their role or one that does not make reference to an actual ecosystem would be awarded 1 or 2 marks.

(c) In the deciduous forest in Breen Wood there are various decomposers such as earthworms, bacteria and fungi. They operate at each trophic level to break down and cause the decay of the organic matter not consumed by other organisms, including excrement and bones. This means that they play an important part in the recycling of the organic nutrients within Breen Wood.

e **4/4 marks awarded**. This full answer provides a spatial context, identifies named decomposers and explains their role.

(d) Explain why trophic pyramids do not have more than 4 or 5 levels. (3 marks)

ℯ 3 marks are awarded for an answer that identifies that energy is lost between trophic levels [1 mark] and gives examples of how this happens (e.g. respiration, excrement, non-consumed elements) [2 marks]. Award 1 mark for an answer that fails to give examples.

> **(d)** The reason why there are seldom more than 4 or 5 trophic levels is that energy is lost as it is passed from one trophic level to the next one.

ℯ **1/3 marks awarded**. The answer identifies the issue of energy loss but fails to provide any examples of how this can occur.

Question 3 Atmosphere: short structured questions

(a) Study Resource 3, which shows some of the responses to Hurricane Floyd in the USA in September 1999. Use the resource and your own case study material to discuss how hurricane protective measures can be used to reduce the loss of life and damage to properties. (6 marks)

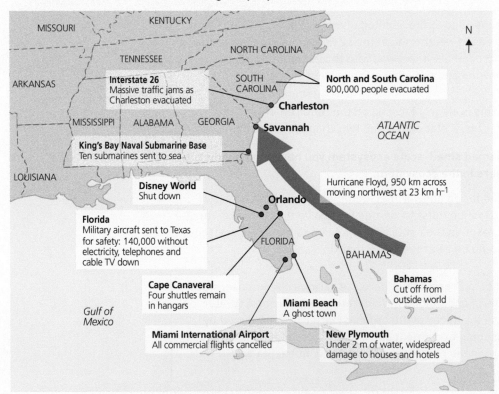

Resource 3 Responses to Hurricane Floyd, September 1999

ℯ A good answer will make reference to some of the features on the resource *and* some case study material that looks at the impact on both people and on property.

Level 3 (5–6 marks): The answer discusses a range of protection measures/ responses, making reference to the resource *and* the student's own hurricane case study material. Discussion should involve how the measures have been able to reduce the loss of life and/or damage to property.

Level 2 (3–4 marks): There is less detail about the measures taken to reduce the death toll and/or damage to property. The case study or resource might be left out at this level.

Level 1 (1–2 marks): A basic answer, which lacks explicit reference to a case study and/or the resource material.

Student answer

(a) When Hurricane Floyd hit the USA in 1999, the authorities were able to instigate an evacuation programme to get people living near the coast and in vulnerable areas to move to places of safety. Over 800,000 people were moved from North and South Carolina, flights were cancelled and Disney World was closed to keep people safe.

The same things happened before Hurricane Katrina hit — on 28 August 2005 the Mayor of New Orleans announced the first evacuation of the city. However, one problem with this evacuation was that many of the people in the city were very poor and had no means of transport to get out of the city.

When Hurricane Floyd hit, some of the naval submarines had been sent out to sea to minimise damage and other aircraft had been flown to Texas and the space shuttles were kept in their hangars. In New Orleans, people were able to board up their properties as they were given a warning from the Early Warning System — though they did not really expect the flooding that took place.

ⓔ 5/6 marks awarded. This makes some use of the resource to answer both aspects of the question: how hurricane protection measures can reduce the loss of life and damage to properties. It also mentions the student's case study material —the impact of Hurricane Katrina on New Orleans — but could have gone into a little more specific detail.

(b) Outline *one* factor that might influence the global temperature pattern. (3 marks)

ⓔ This requires you to choose and outline one of the controls or influences on the global temperature pattern, for example ocean currents, continentality, altitude or prevailing winds.

3 marks are awarded for an answer that develops a detailed description of *one* influence. There should be reference to the differences in global temperature, with places near the poles being much colder than at the Equator. 2 marks are awarded for an answer that goes some way to describe the influence of one factor but which maybe does not explain the global variation in enough depth. An answer that does identify a valid factor but is vague and does not develop specific detail is worth just 1 mark.

(b) Ocean currents can have a big impact on the global temperature pattern. The sea can absorb heat over a long period of time compared with the land and it retains its heat for a much longer time, which means that the ocean currents move blocks of warm water and air towards colder areas. This movement around the globe moves warm water from the Equator towards the poles. A good example is the North Atlantic Current, which moves warm water from the Caribbean to Western Europe. This keeps the UK much warmer in the winter than inland Europe.

e **3/3 marks awarded**. This identifies a valid factor, explaining it in good depth and making reference to the fact that some areas of the Earth are warmer than others due to this factor.

(c) **Describe the differences between horizontal and vertical heat transfer when talking about the global energy balance.** (6 marks)

e A good answer will make reference to more than one difference between the horizontal and vertical methods of heat transfer.

Level 3 (5–6 marks): The answer shows a sound awareness of the main distinctions between horizontal and vertical heat transfer. There is valid reference to the global energy balance.

Level 2 (3–4 marks): There is less detail about the differences. The candidate has maybe only identified and described one difference or has explained two in less detail.

Level 1 (1–2 marks): A basic answer, which lacks explicit reference to detail and does not give specific examples or makes no reference to the global energy balance.

(c) Horizontal heat transfer transfers heat towards or away from the Equator. An example of this is ocean currents or the wind. Ocean currents will move bodies of warm and cold water around the world, which can have a big impact on the climate and weather. Vertical heat transfer transfers heat from ground level into the upper atmosphere or troposphere, for example radiation or convection currents.

e **4/6 marks awarded** This gives good definitions of the two different concepts but needs further development. The answer gets to the top of Level 2 but needs to make more reference to the global energy balance to get into Level 3.

■ Section B

Question 4 Rivers: extended question

With reference to a large-scale drainage basin or delta you have studied, discuss the physical and human causes of flooding.

(15 marks)

_e_ You should name a large-scale drainage basin or delta for Level 3 and discuss a range of both human and physical causes for the flood event or events named. Physical causes might include seasonal climatic change, for example snowmelt or higher-than-average rainfall. Human factors could include deforestation, urbanisation, dam building or farming practices.

Level 3 (10–15 marks): The answer shows a good balance between physical and human factors, describing and explaining a range of causes of flooding. The case study is named and there are good references to case study details throughout. Sound understanding is communicated effectively using specialist terminology.

Level 2 (6–9 marks): The answer describes a range of human and physical factors, but with limited reference to case study detail. Alternatively, the answer is unbalanced, focusing too much on either physical or human. The answer may lack depth and fewer specialist terms are used.

Level 1 (1–5 marks): The answer may cover only one of the two types of cause identified in the question. The answer is superficial with no meaningful case study reference. The level of written communication may also be poor.

Student answer

The Mississippi 2011 floods were caused by a variety of physical and human factors.

The first factors I will look at are physical or climatic factors. The first of these is heavy snowfall in the upper Mississippi and Missouri tributaries to the northwest where snowfall in the 2010/11 winter created a snowpack 60 cm deeper than average. This then melted in the spring and a large volume of water flowed into the rivers, creating a flood peak discharge that reached 15 m at Thebes, which was the largest since records began. At the same time, in the northeast of the drainage basin around the Ohio River tributary, a second physical factor of record-breaking rainfall was in operation. During April, a series of four intense rainstorms produced six times the average monthly rainfall in this area. This also resulted in a flood peak travelling down the river towards Cairo, where the confluence between the Ohio River and the Missouri/upper Mississippi Rivers is located. These two flood peaks met at this confluence, creating a massive flood peak heading into the downstream area of the lower Mississippi Basin.

Once the physical causes had triggered the flood, a variety of human factors increased the scale of the flooding. The first of these was deforestation. Since European settlers first arrived in large numbers in the Mississippi Basin, they have been cutting down trees. In the 1700s, the Lower Mississippi Basin had about 10 million hectares of forest and swamp forest. Today, only about 20% of this remains. Between 1900 and 1935, about 50% was cleared for farming. From 1950 to 1970, another 2 million hectares were cleared. This loss of tree cover increases the flood risk because there is less interception of the rain. This means that it reaches the river channel more quickly, causing higher flood peaks, such as the 15m peak at Thebes, to occur. Fewer trees means less transpiration of water out of the drainage basin. This increases the overall amount of water in the river system.

A second human cause was urbanisation on the floodplains. People know about the flood risk here, but still development is taking place. For instance, Chesterfield Valley in St Louis was an area that was under 3m of flood water during the 1993 flood. But since then, 28,000 new homes have been built here. Furthermore, in the Missouri area, there was urban development on land that had also been under water during the 1993 flood. The greater the proportion of urban development on a floodplain, the greater the risk of flood, as urban areas increase the amount of impermeable surfaces. This results in more overland flow and this, combined with the drains and sewers in urban areas, gets water to the river more quickly, producing faster and higher peak discharges.

ⓔ **14/15 marks awarded** This answer includes a range of physical and human factors (two of each). The case study is named and there is good spatial detail throughout, although one or two more facts would have been preferable. There is a demonstration of good understanding of how the factors link to the flood risk. There is good use of terminology (such as flood peak and confluence) and the quality of written communication is very effective in that the answer clearly demonstrates a depth of understanding of the issues.

Question 5 Ecosystems: extended question

Discuss the processes of environmental change that have resulted in the seral stages in a vegetation succession you have studied at a small or regional scale. (15 marks)

ⓔ This answer must have a clear knowledge of the various seral stages from the succession, and should include references to the plants found at each stage. It must also clearly explain the environmental changes (which may include the soil and the microclimate) that bring about the changes in the plants.

Level 3 (10–15 marks): A vegetation succession is clearly identified and the seral stages outlined in detail with references to the plants found there. A detailed explanation of the changes in environment that drive the succession is given. The answer is well structured and there is very good use of geographical terms.

Level 2 (6–9 marks): An appropriate succession is discussed, and seral stages are recognised. However, the processes of modification of the environment are underdeveloped or lack clarity. The level of written communication is reasonable.

Level 1 (1–5 marks): A basic answer with little detail of either content or explanation. The discussion of the changes in the environment will be absent and the quality of communication is limited, with very little use of geographical terms.

> **Student answer**
>
> The environment changes noticeably as plant succession occurs, as we will see by examining the psammosere succession in the sand dunes at Portstewart Strand.
>
> First, we have the embryo dunes. The main plant here is the sand couch, a hardy plant that can stand the very harsh environment found here. The soil here has very little organic matter, very little moisture content, it contains high salt levels and the pH is very alkaline because of the sea shells in the sand. In addition, the microclimate is harsh as there is no shelter from the north winds that blow here from the Atlantic Ocean.
>
> However, the sand couch begins to modify this harsh environment. Its roots help to stabilise the sand and help the dunes grow in height as the leaves trap more sand. As well as this, more organic matter is added as the plants die and humus content increases.
>
> This changed environment creates conditions suited to plants from the next seral stage: the foredunes. The main plant found here is marram grass. It is adapted to the harsh conditions here as its leaves are rolled tightly and have a shiny surface. This helps it retain moisture as transpiration is reduced. Being buried in sand stimulates its growth and its rhizome root system spreads laterally.
>
> The next seral stage is the dune slack, found over the ridge that runs parallel to the beach. Here, a much wider variety of plant species grows. Initially, this includes mosses, which grow in gaps between the marram grass. In addition, you can find rosette plants such as dandelions growing here too.
>
> After the dune slack, towards the back of the dune system, comes the old, stable grey dunes. Here, the marram grass has increasingly been replaced and other plants such as lady's bedstraw and bird's foot trefoil are found. The dune system is managed by the National Trust and held at plagioclimax, so trees are generally not found here. However, occasionally the odd isolated willow or hazel tree may be found towards the back of the dunes.

ⓔ **8/15 marks awarded**. This answer starts well, as it outlines the characteristics of the first seral stage as well as discussing the harsh environment found there. Paragraphs two and three carry on with this focus of the environmental change, just as the question requires. However, in the rest of the answer the theme of environmental change is missing, and the answer slips into mere description of the seral stages and the plants. Thus it drops down into mid-Level 2.

Question 6 Atmosphere: extended question

(a) Discuss the formation of a mid-latitude depression and make reference to how it works with reference to air masses and the polar front jet stream.   (15 marks)

ⓔ There are three aspects to this question that need to be addressed: you need to show an understanding of the formation processes of a depression and then demonstrate how the formation works in relation to both air masses and the polar front jet stream. Answers should go into depth with clear understanding shown.

Level 3 (10–15 marks): The answer shows detailed understanding of the formation processes. There is clear discussion of the way that a depression forms in the mid-latitudes. The candidate has detailed and accurate knowledge of both the role that air masses and the polar front jet stream plays.

Level 2 (6–9 marks): There is a general understanding of the formation of a depression and some explanation of both air masses and the polar front jet stream. However, depth might be limited. Alternatively, students might only make reference to either air masses or the polar front jet stream.

Level 1 (1–5 marks): A basic answer with little, if any, depth, balance or explanation. Some inaccuracies might be evident.

Student answer

(a) Depressions are one of the main weather systems to affect the British Isles. They are formed out in the North Atlantic Ocean when cold polar maritime air from the north moves south and joins up with some warm tropical maritime air moving north. When the air meets, this creates a front and frontal rainfall is caused. The lighter warm air is forced to rise over the slower moving, more dense cold air. The depression continues to move from the west towards the east and brings low pressure, cloud, rain and windy conditions to the UK. Depressions are formed in the mid-latitudes when air from two air masses meet. Over the Atlantic Ocean, cold air from the north (the polar maritime air mass) comes down and produces unstable air, which then meets up with the warmer air from the south (the tropical maritime air mass). As the air moves north, the less dense warm air from the south will start to rise up over the colder and heavier air from the north. This creates the warm front and can cause cloud and precipitation to form.

ⓔ **9/15 marks awarded.** This student has presented some good information and development of the facts of the particular case study, while addressing the two elements of the question. There is good, detailed discussion of the formation processes and also some development of the role that air masses have played in the formation of the depression. However, no detail was made in reference to the polar front jet stream and its role in forming and directing depressions in the Atlantic. There could have been a little more specific information added here to take the answer into Level 3.

(b) With reference to your case study of a hurricane, evaluate the management strategies used to reduce its impact on people and property.   (15 marks)

e You must name a specific hurricane event and then evaluate the particular management strategies that were put in place to reduce its impact on people and property. Evaluating demands more than a simple discussion — there should be reference to strengths and weaknesses of measures and some critical reflection.

Level 3 (10–15 marks): The answer shows detailed understanding of the hurricane and the management strategies used are outlined and evaluated. There is good depth and explanation of impact on people and on property.

Level 2 (6–9 marks): There is a general understanding of the case study. The level of knowledge of the case study might be limited. There might be some attempt to evaluate protective strategies. Alternatively, the impact on people or property might be ignored.

Level 1 (1–5 marks): A basic answer with little, if any, case study material.

(b) On 29 August 2005, Hurricane Katrina hit the city of New Orleans on the southern coast of the USA. The hurricane brought high winds and rainfall and created a storm surge, which burst the banks of the levées that protected the city from flooding. After the flood waters had gone down, many of the people who lived in the area were critical of the protective measures that had been taken to reduce the loss of life and damage to property.

Measures taken to reduce the impact on people

Positive: In the USA there are organisations like the National Hurricane Centre that will track hurricanes and work out where they are going to make landfall. This means that only people who really need to be evacuated will be moved. New Orleans did have its own disaster plan, which had been practised many times. The emergency services had planned for a disaster like Katrina so they should have been ready for whatever happened.

Negative: FEMA — the national USA organisation for managing disasters — reacted slowly to the disaster. It did not take decisions quickly enough in the early stages of the emergency. Even though the plans had been rehearsed, many of the poor people in the centre of the city were left to themselves as they had no way of getting out of the city. Up to 25% of the population had no car, plus the final decision to evacuate the city was made only 19 hours before the hurricane arrived. This meant that there was not enough time to get everyone out of the city.

Measures taken to reduce the impact on property

Positive: Most of New Orleans is below sea level. US army-built levées were designed to protect the city in the event of a hurricane or flood. However, these were not built strong or high enough for the impact of Katrina — a category 5 hurricane — though many people note that things would have been a lot worse if these measures were not taken at all.

Negative: The levées were ruptured very quickly and this caused a mass of flood water to move through the city. Buildings were not designed to withstand both the high winds and the high water levels. The government was more concerned with saving lives; they did not care that property was being damaged.

ⓔ **11/15 marks awarded**. There is some good development and discussion about the management strategies used, with reference to Hurricane Katrina and New Orleans. There is a nice balance between the positive and negative and between property and people. However, more detail could have been given in relation to property, especially the negative impacts. It would have been good to see more facts and figures throughout the discussion as well. It is a Level 3 answer but still could have developed more detail.

Knowledge check answers

1 Infiltration refers to water entering the ground surface and transferring into the soil store, whereas percolation refers to the movement of water down through the soil profile until it reaches the groundwater store.

2 During the summer, trees have leaves, which increases the store of interception. This delays the water reaching the ground by the transfer of stem flow and so the ground is less likely to have its infiltration capacity exceeded. This results in less overland flow. In contrast, during the winter there is less interception and so the water arrives at the ground more quickly and more overland flow results.

3 As a drainage basin experiences increased urbanisation, lag time is reduced and peak discharge increases. This is due to the increase in impermeable surfaces and the introduction of drains and sewers. These both increase overland flow and provide an efficient route for the water to get to the river, and so the water arrives at the river more quickly.

4 Corrasion/abrasion, hydraulic action and attrition are affected by discharge and energy levels as they are physical processes. If a river has more energy, these processes are more efficient. In contrast, corrosion is a chemical reaction between weak acids in the river water and the rock over which the river flows.

5 In cross-section, meanders are asymmetrical. On the outside of the bend they have a steeper bank (which is often undercut by erosion) known as a river cliff. On the inside of the bend, they have a much more gently sloping bank of point bar deposits, known as a slip-off slope.

6 The rate can be affected by how steep the sea floor gradient is. The steeper the gradient, the slower the rate of formation, as the sediment being deposited is more likely to fall down the slope under the influence of gravity.

7 Re-sectioning seeks to improve both channel capacity (by increasing channel cross-sectional area) and efficiency (by reducing friction due to water contact with the bed and banks, and by increasing river velocity). On the other hand, realignment improves channel efficiency by removing meanders and straightening the channel. As a result, the river drops by the same height as before, but over a shorter distance. This means that the gradient becomes steeper, increasing velocity.

8 I agree to a certain extent. When flooding is at a large scale, it can bring many devastating consequences including loss of life, displacement of people from their homes, damage to property and infrastructure, and loss of crops. However, smaller scale flooding — and even larger scale flooding, in the longer term — can bring some benefits including promoting habitat diversity, an increase in soil groundwater store for agriculture and for wells for human use, increased fish stocks (to improve diets, especially in less economically developed countries (LEDCs)) and providing recreational opportunities (e.g. wetland fishing) that can boost local economies.

9 Both temperate grasslands and tundra areas have cold winters with temperatures below −30°C. However, during the summer, temperatures in the grasslands can exceed 35°C whereas in the tundra, temperatures struggle to reach 10°C. Precipitation totals in the temperate grasslands are low overall at around 400–500 mm per year whereas in the tundra they are very low, often only around 100 mm per year. In addition, the hot summer temperatures of the temperate grasslands result in high amounts of evapotranspiration — something that is much more limited in the cool summers of the tundra.

10 Trophic level 2 consists of the herbivores. These animals feed on the plants and so are called heterotrophs or primary consumers. At trophic level 3 we find the carnivores that feed on the herbivores from trophic level 2, so they are heterotrophs, or secondary consumers.

11 Energy flows, i.e. it enters via photosynthesis, moves through the trophic levels and is finally outputted as heat via respiration. In contrast, nutrients cycle, i.e. they move between the stores (biomass, litter and soil) within the ecosystem.

12 The three aspects that change are vegetation, soils and microclimate.

13 In the psammosere at Portstewart Strand, the soil quality improves considerably as the pH changes from being very alkaline in the embryo dunes to around 6.5 in the grey dunes. Species diversity increases from mostly sand couch in the embryo dunes to a range of stratified plants in the grey dunes, including bedstraw and bird's foot trefoil.

14 The three processes that can transfer heat vertically in the atmosphere are radiation (where the land radiates heat back out to space through long-wave radiation), convection (where warm air is forced to rise as part of convection currents, and rising warm air is replaced by colder, descending air) and conduction (where the energy is transferred through contact).

15 Heat energy is circulated from the Equator to the poles according the tri-cellular model. The sun directly heats the air at the Equator causing the warm air to rise and diverge towards the poles (Hadley cell). This air starts to sink again between 30° N and S of the Equator and moves away from the high pressure towards the 60° N and S point (Ferrel cell) where the air starts to rise again at the polar front (Polar cell).

16 Continentality influences the pattern of temperature because areas that are further inland experience a much bigger range of temperatures than those on the coast. It is generally warmer in the summer and colder in the winter inland.

17 The Coriolis effect deflects the flow of air in the northern hemisphere to the right, while in the southern hemisphere it deflects it to the left.

18

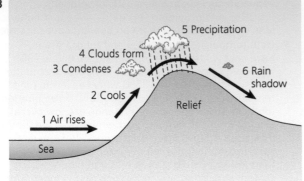

Orographic/relief rainfall

Air is forced to rise over a physical barrier such as mountains. As the air is forced to rise, it cools, condenses, clouds form and precipitation happens at the top of the hill.

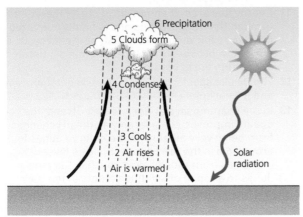

Convectional rainfall

The direct heating of the sun warms air above the surface of the Earth. This causes the air to rise, then it cools, condenses, and clouds form, leading to precipitation.

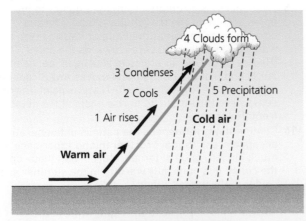

Frontal/cyclonic rainfall

This happens when two bodies of air (air masses) meet. They meet at a front and the air becomes unstable as warm and cold air cannot mix, so the warm air starts to rise up over the colder dense air. As this happens, the air cools, condenses, then clouds form and precipitation takes place.

19 The main air mass that affects the British Isles is the tropical maritime air mass. This brings warm and wet conditions and causes dull skies (nimbostratus clouds), drizzle and fog.

20 A depression is an area of low atmospheric pressure that can produce cloudy, rainy and windy weather conditions. Depressions are formed when cold polar maritime and warm tropical maritime air meet in the Atlantic Ocean, creating a weather front. The lighter, warm air starts to rise up over the colder and denser air from the north.

21 As a depression starts to pass overhead the first thing people notice is that the pressure starts to fall and temperatures start to decrease. There is an increase in cloud cover and in wind speeds. As the warm front passes overhead the pressure is low and there is more cloud, with some periods of precipitation. Temperatures increase in the warm sector that follows and skies clear to some extent as calmer conditions develop and less rain falls. However, very quickly temperatures fall as the cold front moves into position, bringing much deeper clouds, as wind speeds increase dramatically and the amount and intensity of rainfall increase. After the cold front has passed conditions remain cold but clouds gradually break up, and wind speeds and chances of further precipitation decrease.

22 The main differences between a summer and winter anticyclone are mostly to do with temperature and cloud cover, and the influences that these bring. In the summer the lack of cloud cover allows temperatures to build, whereas in the winter the lack of cloud cover means that days can be cold, while at night the temperatures can go below freezing, causing frost and icy conditions. The main similarities are that anticyclones usually bring periods of dry, calm, stable weather with few clouds and little precipitation.

23 The main differences are that a warm front is the leading edge of a block of warm air, a cold front leads a block of cold air and an occluded front will have cold air at the bottom and warmer air higher up.

24 Hurricanes usually need to form near the Equator, in areas where the seawater temperature is above 27°C. The airflow needs to start with convectional currents, which allow air to rise and cool quickly in the high levels of the atmosphere. The relative humidity of the air needs to be high so that a massive amount of moisture can be dealt with quickly.

Note: **Bold** page numbers indicate defined terms.

Index